SpringerBriefs in Statistics

JSS Research Series in Statistics

The current research of statistics in Japan has expanded in several directions in line with recent trends in academic activities in the area of statistics and statistical sciences over the globe. The core of these research activities in statistics in Japan has been the Japan Statistical Society (JSS). This society, the oldest and largest academic organization for statistics in Japan, was founded in 1931 by a handful of pioneer statisticians and economists and now has a history of about 90 years. Many distinguished scholars have been members, including the influential statistician Hirotugu Akaike, who was a past president of JSS, and the notable mathematician Kiyosi Itô, who was an earlier member of the Institute of Statistical Mathematics (ISM), which has been a closely related organization since the establishment of ISM. The society has two academic journals: the *Japanese Journal of Statistics and Data Science* (JJSD, Springer), which is the successor of the *Journal of the Japan Statistical Society* (JJSS) and the *Journal of the Japan Statistical Society* (Japanese Series). The membership of JSS consists of researchers, teachers, and professional statisticians in many different fields including mathematics, statistics, engineering, medical sciences, government statistics, economics, business, psychology, education, and many other natural, biological, and social sciences. The JSS Series of Statistics aims to publish recent results of current research activities in the areas of statistics and statistical sciences in Japan that otherwise would not be available in English; they are complementary to the two JSS academic journals, both English and Japanese. Because the scope of a research paper in academic journals inevitably has become narrowly focused and condensed in recent years, this series is intended to fill the gap between academic research activities and the form of a single academic paper. The series will be of great interest to a wide audience of researchers, teachers, professional statisticians, and graduate students in many countries who are interested in statistics and statistical sciences, in statistical theory, and in various areas of statistical applications.

Nobuhiro Taneichi · Yuri Sekiya

Improving Tests for Discrete Small Sample Data

Nobuhiro Taneichi
Sapporo, Hokkaido, Japan

Yuri Sekiya
Kushiro, Hokkaido, Japan

ISSN 2191-544X ISSN 2191-5458 (electronic)
SpringerBriefs in Statistics
ISSN 2364-0057 ISSN 2364-0065 (electronic)
JSS Research Series in Statistics
ISBN 978-981-95-5300-6 ISBN 978-981-95-5301-3 (eBook)
https://doi.org/10.1007/978-981-95-5301-3

This Springer imprint is published by the registered company Springer Nature Singapore Pte Ltd.
The registered company address is: 152 Beach Road, #21-01/04 Gateway East, Singapore 189721, Singapore

Preface

This book considers test statistics that are based on ϕ-divergence for discrete multivariate models. Concretely, we consider a multinomial goodness-of-fit test and a test of independence of contingency tables based on ϕ-divergence when the sample size is not so large. When sample size is not large enough, approximation of the distribution of test statistic by a chi-square distribution is not good. Therefore, there is a risk that the results of a test will lead to the opposite conclusion. This book is composed of two parts. In the first part, by using the expression of an asymptotic expansion of the original statistic, we construct a transformed statistic whose distribution is closer to a chi-square distribution. A transformed statistic for a 2-way independence model of a contingency table (Chap. 3), a complete independence model, an independence model among a group of factors in an M-way contingency table (Chap. 4), a conditional independence model (Chap. 5), and generalized linear models of binary data (Chap. 6) are given. Furthermore, a transformed log likelihood ratio statistic for loglinear models is given (Chap. 7). In the second part, the selection of statistics for which the distribution is close to the limiting distribution is discussed using the evaluation of second-order correction terms (Chaps. 8 and 9).

The book is aimed at applied statisticians, postgraduate students in statistics and statistically sophisticated researchers.

Detailed proofs of the results are not provided in each chapter. So, important proofs of the results are shown in the Appendix.

We are very grateful to Masafumi Akahira, emeritus professor of University of Tsukuba who recommended us the writing of this book. We are also grateful to Yutaka Hirachi of Springer Nature Japan who cooperated for writing of it.

Sapporo, Japan | Nobuhiro Taneichi
Kushiro, Japan | Yuri Sekiya

Competing Interests The authors have no competing interests to declare that are relevant to the content of this manuscript.

Contents

Chapter 1
Purpose of This Book

Keywords Asymptotic test · Categorical data · Contingency table · Exact methods · Monte Carlo simulation · Sample size

1.1 Analysis of Categorical Data When the Sample Size Is Not So Large

Categorical data analysis are used in various field, for example, psychology, sociology, economics, biology, medicine, etc. There are many situations in which we cannot obtain enough data for categorical data analysis based on large sample theory. In such situations, this book proposes an improvement of the test statistics and conditions of selecting the test statistics for asymptotic tests.

We consider the data shown in Table 1.1 to explain the motive for this book. The data are from an investigation conducted at the Danish National Institute for Social Research in Copenhagen quoted in Edwards and Kreiner (1983). We extract a 4-way contingency table from the original 5-way contingency table. In the table, 390 employed men in the age group of 18–67 years are cross-classified according to whether they are skilled workers, unskilled workers, or office workers (Employment: Skilled, Unskilled, Office), whether they live in a rented apartment or an owned apartment (Mode: Rent, Own) and whether or not (Response: Yes, No) they had done any work in the preceding 12 months that they would previously have paid a craftsman to do. One of the purposes of the investigation, for which data were collected in 1978–79, was to estimate the extent of tax evasion in the construction industry.

To test the hypothesis that age, employment, and pair (Response, Mode) are independent in the data in Table 1.1, the following procedure would usually be carried out. First, the value of the log likelihood ratio test statistic G^2 would be computed. The statistic is a special case of (4.3) in Sect. 4.1. Here, the observed value of G^2 is

N. Taneichi and Y. Sekiya, *Improving Tests for Discrete Small Sample Data*,
JSS Research Series in Statistics,
https://doi.org/10.1007/978-981-95-5301-3_1

Table 1.1 A 4-way table extracted from Edwards and Kreiner (1983)

			Age		
Employment	Mode	Response	18–30	31–45	46–67
Skilled	Rent	Yes	18	15	6
		No	15	13	9
	Own	Yes	5	3	1
		No	1	1	1
Unskilled	Rent	Yes	17	10	15
		No	34	17	19
	Own	Yes	2	0	3
		No	3	2	0
Office	Rent	Yes	30	23	21
		No	25	19	40
	Own	Yes	8	5	1
		No	4	2	2

41.449. Since G^2 is asymptotically distributed as χ^2_{28}, the hypothesis is rejected at the asymptotic 5% level given that $\chi^2_{28}(0.05) = 41.337$. The asymptotic p-value for G^2 is 0.04882 (Details are explained in Sect. 4.5). Such a conclusion is only valid on the basis of the assumption that the distribution of the test statistic G^2 is suitably approximated by a chi-square distribution with appropriate degrees of freedom. If the asymptotically approximated critical point $\chi^2_f(\alpha)$ and approximate p-value based on χ^2_f are inaccurate, there is a risk that the results of the approximated test stated above will lead to the opposite conclusion of the test given by an exact critical point. In order to avoid such a risk, exact methods based on direct enumeration of probability are considered (Agresti 1992; Storer and Kim 1990; Tang and Tang 2002). Methods based on Monte Carlo simulation such as the Markov chain Monte Carlo method (Diaconis and Efron 1985; Ohkubo 2000; Rapallo 2003) and the bootstrap method (Jeong et al. 2005; Amiri and von Rosen 2011) are also considered. However, computation using an exact method is very difficult when the number of dimensions of the contingency table and sample size are not small. Monte Carlo methods are effective when the number of dimensions of the contingency table is small. However, as the sample size and number of dimensions of the contingency table increase, the amount of computation for Monte Carlo methods becomes enormous.

1.2 Contents of This Book

This book is composed of the following two parts. In the first part, for various kinds of independence models for a contingency tables and a generalized linear model for binary data, we propose a test procedure for obtaining more reliable results compared to when using an asymptotic distribution of the test statistic without an enormous amount of computation when the sample size is not so large. For the test procedure, we use a transformed test statistic. The transformed statistic is constructed by an approximation of the distribution of test statistics based on an asymptotic expansion that is more accurate than an asymptotic approximation. In Chap. 2, Preliminary results for the first part are introduced. In Chap. 3, transformed statistics for independence in a 2-way contingency table are derived. In Chap. 4. transformed statistics for the test of complete independence and statistics for the test of independence among groups of factors are derived. In Chap. 5, transformed statistics for conditional independence in a 3-way contingency table are derived. In Chap. 6, transformed goodness-of-fit test statistics for a generalized linear model of binary data are derived. In Chap. 7, a transformed log likelihood ratio test statistic for a loglinear model for a contingency table is derived.

In the second part of the book, for a complete independence model for a multi-way contingency table, we derive a condition for a test statistic whose distribution is close to a chi-square distribution, which is an asymptotic distribution, even when the sample size is not so large. In Chap. 8, the second order correction term is considered as a criterion to derive the condition. We show that only Pearson's X^2 statistic satisfies the condition among well-known families of statistics. Furthermore, in Chap. 9, we construct a new statistic that, in addition to Pearson's X^2 statistic, satisfies the condition.

References

Agresti A (1992) A survey of exact inference for contingency tables. Stat Sci 7(1):131–153

Amiri S, von Rosen D (2011) On the efficiency of bootstrap method into the analysis contingency table. Comput Meth Prog Bio 104:182–187

Diaconis P, Efron B (1985) Testing for independence in a two-way table: new interpretations of the chi-square statistic. Ann Stat 13(3):845–874

Edwards D, Kreiner S (1983) The analysis of contingency tables by graphical models. Biometrika 70:553–565

Jeong HC, Jhun M, Kim D (2005) Bootstrap tests for independence in two-way ordinal contingency tables. Comput Stat Data Anal 48:623–631

Ohkubo N (2000) MCMC method for $I \times J$ contingency tables: extended Mantel test and correlation test (in Japanese). Jpn J Applied Statistics 29(3):131–139

Rapallo F (2003) Algebraic Markov bases and MCMC for two-way contingency tables. Scand J Stat 30(2):385–397

Storer BE, Kim C (1990) Exact properties of some exact test statistics for comparing two binomial proportions. J Am Stat Assoc 85:146–155

Tang NS, Tang ML (2002) Exact unconditional inference for risk ratio in a correlated 2×2 table with structural zero. Biometrics 58:972–980

Chapter 2
Preliminary Results

Keywords Asymptotic expansion · Bartlett-type adjustment · Edgeworth expansion · Improved transformation · ϕ-divergence · Power divergence · The Bartlett adjustment · Transformed statistics

2.1 Test Statistic Based on a ϕ-Divergence for a Multinomial Model for a Simple Null Hypothesis

We introduce the family of test statistics based on ϕ-divergence for a multinomial model as follows.

A measure for dissimilarity between distributions called ϕ-divergence was proposed independently by Csiszár (1967) and Ali and Silvey (1966). The ϕ-divergence measure between discrete distributions $p = (p_1, \ldots, p_k)^\top$ and $q = (q_1, \ldots, q_k)^\top$ is defined by

$$D_\phi(p, q) = \sum_{j=1}^{k} q_j \phi\left(\frac{p_j}{q_j}\right),$$

where $\phi(t)$ is a real convex function for $t > 0$ with

$$\phi(0) = \lim_{t\to+0} \phi(t), 0\phi\left(\frac{x}{0}\right) = \lim_{y\to+0} y\phi\left(\frac{x}{y}\right) = x \lim_{t\to+0} t\phi\left(\frac{1}{t}\right), \text{ if } x > 0,$$

and $0\phi\left(\frac{0}{0}\right) = 0$.

This chapter is based on the following article: Journal of the Japan Statistical Society 31(2), Taneichi N., Sekiya Y. and Suzukawa A., An asymptotic approximation for the distribution of ϕ-divergence multinomial goodness-of-fit statistic under local alternatives, 207–224, (2001).

N. Taneichi and Y. Sekiya, *Improving Tests for Discrete Small Sample Data*,
JSS Research Series in Statistics,
https://doi.org/10.1007/978-981-95-5301-3_2

Let $X = (X_1, \ldots, X_k)^\top$ be distributed according to a multinomial distribution $\mathcal{M}_k(n, p)$, where $\sum_{j=1}^k X_j = n$, $\sum_{j=1}^k p_j = 1, 0 < p_j < 1, j = 1, \ldots, k$, and $p = (p_1, \ldots, p_k)^\top$ is an unknown probability vector. In order to test the null hypothesis

$$H_0 : p = q \tag{2.1}$$

for some completely specified probability vector $q = (q_1, \ldots, q_k)^\top$, ϕ-divergence statistic C_ϕ was introduced by Zografos et al. (1990). C_ϕ is defined by

$$C_\phi = 2n D_\phi \left(\frac{X}{n}, q \right), \tag{2.2}$$

where ϕ is a function satisfying $\phi(1) = \phi'(1) = 0$ and $\phi''(1) = 1$ (Pardo et al. 1999).

2.2 Special Families of ϕ-Divergence Statistics

When we choose the following convex function ϕ^a as ϕ, the ϕ-divergence measure becomes the power divergence measure (Read and Cressie 1988).

$$\phi^a(t) = \begin{cases} \{a(a+1)\}^{-1}\{t^{a+1} - t + a(1-t)\} & (a \neq 0, -1) \\ t \ln t + 1 - t & (a = 0) \\ -\ln t - 1 + t & (a = -1). \end{cases} \tag{2.3}$$

Therefore, the ϕ-divergence family of statistics C_ϕ includes the power divergence family of statistics $R^a \equiv C_{\phi^a}$ proposed by Cressie and Read (1984) and Read and Cressie (1988). It is well known that the family of statistics R^a includes Pearson's X^2 statistic ($a = 1$), the log likelihood ratio statistic ($a = 0$), the Freeman-Tukey statistic ($a = -1/2$), the modified log likelihood ratio statistic ($a = -1$), the Neyman modified X^2 statistic ($a = -2$), and the statistic recommended by Cressie and Read (1984) and Read and Cressie (1988) ($a = 2/3$). Rukhin (1994) considered the family of statistics given by the following convex function ϕ_Q^a as ϕ:

$$\phi_Q^a(t) = \frac{(t-1)^2}{2\{a + t(1-a)\}}, \quad a \in [0, 1]. \tag{2.4}$$

The family of statistics $Q^a \equiv C_{\phi_Q^a}$ includes Pearson's X^2 statistic ($a = 1$) and the Neyman modified X^2 statistic ($a = 0$). Furthermore, Lin (1991) considered the family of divergences given by the following convex function ϕ_L^a as ϕ.

$$\phi_L^a(t) = \{a(1-a)\}^{-1}\{at \ln t - (at + 1 - a)\ln(at + 1 - a)\}, \quad a \in [0, 1].$$

The family of statistics $L^a \equiv C_{\phi_L^a}$ includes the loglikelihood ratio statistic ($a = 0$) and the modified loglikelihood ratio statistic ($a = 1$) by considering continuity in a.

Zografos et al. (1990) showed that under the null hypothesis H_0, all of the members of the ϕ-divergence family of statistics C_ϕ defined by (2.2) are asymptotically distributed according to a chi-square distribution with degrees of freedom $\lambda = k - 1$.

2.3 Test Statistic Based on ϕ-Divergence for Multinomial Model for Composite Null Hypotheses

When we consider testing a model with unspecified parameters, that is, testing composite null hypotheses, the hypotheses are

$$H_0^{\Theta} : p = p(\theta_0) \in \Lambda \text{ for some unknown } \theta_0 \in \Theta,$$

where $\Lambda = \left\{p(\theta) = (p_1(\theta), \ldots, p_k(\theta))^{\top} \in \Delta_k : \theta = (\theta_1, \ldots, \theta_r)^{\top} \in \Theta\right\}$, Θ is an open subset of R_r $(r < k - 1)$, and $\Delta_k = \{p = (p_1, \ldots, p_k)^{\top} : p_j \geq 0,\ j = 1, \ldots, k,\ \sum_{j=1}^{k} p_j = 1\}$. Let ϕ_1 and ϕ_2 be arbitrary concrete functions that satisfy the condition of ϕ-divergence stated above. Also, assuming the regularity condition of Birch (1964) and assuming $p : \Theta \to \Delta_k$ has continuous second partial derivatives in a neighborhood of θ_0. Then under H_0^{Θ}, the ϕ-divergence statistic

$$2nD_{\phi_1}\left(\frac{X}{n}, p(\hat{\theta}_{\phi_2})\right)$$

is asymptotically distributed according to a chi-square distribution with degrees of freedom $k - r - 1$, where $\hat{\theta}_\phi$ is the minimum ϕ-divergence estimator of θ defined by

$$\hat{\theta}_\phi = \arg \inf_{\theta \in \Theta \subset R_r} D_\phi\left(\frac{X}{n}, p(\theta)\right).$$

This result was shown in Pardo (2006). The minimum ϕ-divergence estimator includes the maximum likelihood estimator as in the case of $\phi = \phi^0$, where ϕ^a is given by (2.3).

2.4 The Closer Approximation of Distribution of Multinomial Goodness-of-Fit Test Statistics

We consider approximation of the distribution of multinomial goodness-of-fit test statistics based on an asymptotic expansion under H_0. First, we state the progress of the theory historically as follows:

By using the Euler-Mclaurin summation formula, Esséen (1945) derived an asymptotic expansion for the distribution function of the sum of i.i.d. lattice-valued random variables, with the first term after the normal given by a simple discontinuous function. In the case of higher dimensions, Esséen (1945) also pointed out that the problem of improving the normal approximation for probabilities assigned to spheres or ellipsoids is tied up with the lattice point problem. Ranga Rao (1961) proposed a multidimensional extension of the Euler-Mclaurin summation formula. Using the extended formula, he derived an asymptotic expansion for the probability that the normalized sum of i.i.d. lattice-valued random vectors is contained in a Borel set B. The details are shown in Bhattacharya and Ranga Rao (1976). Using what Esséen pointed out and Ranga Rao's expansion, Yarnold (1972) obtained an expansion that includes a term for the number of lattice points inside the region of interest when B is an extended convex set. Yarnold (1972) also applied it to the null distribution of Pearson's X^2 statistic for goodness-of-fit tests for a multinomial distribution and derived an asymptotic expansion for the null distribution of Pearson's X^2 statistic. The expansion consists of continuous and discontinuous terms. In a fashion similar to the X^2 statistic, approximations based on asymptotic expansions for null distributions of the log likelihood ratio test statistic and the Freeman–Tukey statistic were obtained by Siotani and Fujikoshi (1984). An approximation of power divergence statistics was obtained by Read[a] (1984) and an approximation of ϕ-divergence statistics was obtained by Menéndez et al. (1997).

2.5 Summary of Results of the Closer Approximation

Here, we summarize the results of an asymptotic expansion for a distribution of multinomial goodness-of-fit test statistics based on ϕ-divergence under H_0 given by (2.1). Let $X = (X_1, \ldots, X_k)^\top$ be distributed according to a multinomial distribution $\mathcal{M}_k(n, q)$. We put $r = k - 1$. Let

$$U_j = n^{-1/2}(X_j - nq_j), \quad j = 1, \ldots, k$$

and $U = (U_1, \ldots, U_r)^\top$. Then U is a lattice random vector that takes values in the set

$$L = \left\{u = (u_1, \ldots, u_r)^\top : u = n^{-1/2}(\tilde{x} - n\tilde{q}),\ \tilde{x} \in M\right\},$$

where

$$M = \left\{\tilde{x} = (x_1, \ldots, x_r)^\top : x_1, \ldots, x_r \text{ are non-negative integers and } \sum_{j=1}^{r} x_j \leq n\right\}$$

and $\tilde{q} = (q_1, \ldots, q_r)^\top$. We need the following lemma, which states a local Edgeworth expansion under H_0 given by (2.1).

Lemma 2.1 *Let* $u = n^{-1/2}(\tilde{x} - n\tilde{q})$. *Then for any* $\tilde{x} \in M$,

$$\Pr\{U = u|H_0\} = n^{-r/2} f_0(u) \left\{1 + n^{-1/2} h_1(u) + n^{-1} h_2(u) + O(n^{-3/2})\right\},$$

where

$$f_0(u) = (2\pi)^{-r/2} \mid \Omega \mid^{-1/2} \exp\left\{-\frac{1}{2} u^\top \Omega^{-1} u\right\},$$

$$\Omega = diag(q_1, \ldots, q_r) - \tilde{q}\tilde{q}^\top,$$

$$h_1(u) = -\frac{1}{2}\sum_{j=1}^{k} \frac{u_j}{q_j} + \frac{1}{6}\sum_{j=1}^{k} \frac{u_j^3}{q_j^2},$$

$$h_2(u) = \frac{1}{2}\{h_1(u)\}^2 + \frac{1}{12}(1 - S) + \frac{1}{4}\sum_{j=1}^{k} \frac{u_j^2}{q_j^2} - \frac{1}{12}\sum_{j=1}^{k} \frac{u_j^4}{q_j^3},$$

$$u_k = -\sum_{j=1}^{r} u_j,$$

and

$$S = \sum_{j=1}^{k} q_j^{-1}.$$

By using Theorem 2.1 of Bhattacharya and Ranga Rao (1976), this lemma was obtained by Siotani and Fujikoshi (1984). Yarnold (1972) gave an asymptotic expansion for $\Pr\{U \in G|H_0\}$ when G is an extended convex set. A set G is called an extended convex set if G has the representation

$$G = \{u = (u_1, \ldots, u_r)^\top : \alpha_j(u_j^*) < u_j < \beta_j(u_j^*),$$
$$u_j^* = (u_1, \ldots, u_{j-1}, u_{j+1}, \ldots, u_r)^\top \in G_j\}$$

for every $j \in \{1, 2, \ldots, r\}$, where $G_j \subset R_{r-1}$ and α_j, β_j are continuous functions on R_{r-1} (Yarnold 1972). By using the notation of Siotani and Fujikoshi (1984), Yarnold's result is described as follows:

Theorem 2.1 *If G is a bounded extended convex set, then*

$$\Pr\{U \in G|H_0\} = K_1 + K_2 + K_3 + O(n^{-3/2}), \tag{2.5}$$

where

$$K_1 = \int \cdots \int_G f_0(u) \left\{1 + n^{-1/2} h_1(u) + n^{-1} h_2(u)\right\} du_1 \cdots du_r,$$

$$K_2 = -n^{-1/2} \sum_{j=1}^{r} n^{-(r-j)/2} \sum_{u_{j+1} \in L_{j+1}} \cdots \sum_{u_r \in L_r} \\ \times \int \cdots \int_{G_j} [S_1(\sqrt{n} u_j + n q_j) f_0(u)]_{\alpha_j(u_j^*)}^{\beta_j(u_j^*)} du_1 \cdots du_{j-1},$$

$$K_3 = n^{-1} \sum_{j=1}^{r} n^{-(r-j)/2} \sum_{u_{j+1} \in L_{j+1}} \cdots \sum_{u_r \in L_r} \int \cdots \int_{G_j} \Bigg[-S_1(\sqrt{n} u_j + n q_j) h_1(u) f_0(u) \\ + S_2(\sqrt{n} u_j + n q_j) \frac{\partial}{\partial u_j} f_0(u) \Bigg]_{\alpha_j(u_j^*)}^{\beta_j(u_j^*)} du_1 \cdots du_{j-1},$$

$$S_1(t) = t - [t] - \frac{1}{2},$$

S_2 is the real-valued periodic function with period one such that

$$S_2(t) = \frac{1}{2}\left(t^2 - t + \frac{1}{6}\right) \quad for\ t \in [0, 1],$$

$$[h(u)]_{\alpha_j(u_j^*)}^{\beta_j(u_j^*)} = h(u_1, \ldots, u_{j-1}, \beta_j(u_j^*), u_{j+1}, \ldots, u_r) \\ - h(u_1, \ldots, u_{j-1}, \alpha_j(u_j^*), u_{j+1}, \ldots, u_r),$$

$$L_j = \left\{u_j : u_j = n^{-1/2}(x_j - n q_j) \text{ and } x_j \text{ is integer}\right\},$$

and f_0 and h_j, $j = 1, 2$ are given in Lemma 2.1. Furthermore, it holds that $K_2 = O(n^{-1/2})$ and $K_3 = O(n^{-1})$.

The K_1 term can be regarded as the Edgeworth expansion for a continuous distribution. The K_2 term is a discontinuous term to account for the discontinuity in U. Using Lemma 2.1 and Theorem 2.1, we can obtain an asymptotic expansion of the null distribution of C_ϕ. Let

$$G_\eta^\phi = \{u = (u_1, \ldots, u_r)^\top : C_\phi(u) < \eta\},$$

where

$$C_\phi(u) = 2n \sum_{j=1}^{k} q_j \phi\left(1 + \frac{u_j}{\sqrt{n} q_j}\right).$$

Since the set G_η^ϕ is a bounded extended convex set, we can write $\Pr\{C_\phi(U) < \eta | H_0\} = \Pr\{U \in G_\eta^\phi | H_0\}$ as the formula (2.5) with $G = G_\eta^\phi$.

Theorem 2.2 *If ϕ is four times continuously differentiable, then K_1 and K_2 in the case of $G = G_\eta^\phi$ are evaluated as follows:*

$$K_1 = \Pr\{\chi_r^2 < \eta\} + \frac{1}{24} n^{-1} \sum_{j=0}^{3} r_j' \Pr\{\chi_{r+2j}^2 < \eta\} + O(n^{-3/2}), \tag{2.6}$$

where

$$\begin{aligned}
r_0' &= 2(1 - S), \\
r_1' &= 3(-k^2 - 2k + 3S) + 6\phi'''(1)(-k^2 + S) \\
&\quad + 3\phi^{(4)}(1)(2k - 1 - S) + (\phi'''(1))^2(-3k^2 - 6k + 4 + 5S), \\
r_2' &= 6(k^2 + 2k - 1 - 2S) + 4\phi'''(1)(3k^2 + 3k - 2 - 4S) \\
&\quad + 3\phi^{(4)}(1)(-2k + 1 + S) + 2(\phi'''(1))^2(3k^2 + 6k - 4 - 5S), \\
r_3' &= (1 + \phi'''(1))^2(-3k^2 - 6k + 4 + 5S),
\end{aligned}$$

S is defined in Lemma 2.1, and χ_r^2 denotes a central chi-square random variable with degrees of freedom r. K_2 can be approximated to the first order by

$$\hat{K}_2 = \frac{\{N(\eta) - n^{r/2} V(\eta)\} e^{-\eta/2}}{\left\{ (2\pi n)^r \prod_{j=1}^{k} q_j \right\}^{1/2}}, \tag{2.7}$$

where

$$\begin{aligned}
&V(\eta) \\
&= V_0 \left[1 + \frac{\eta\{(\phi'''(1))^2(5S - 3k^2 - 6k + 4) - 3\phi^{(4)}(1)(S - 2k + 1)\}}{24n(k + 1)} \right] + O(n^{-3/2}),
\end{aligned}$$

$$V_0 = \frac{\left\{ (\pi\eta)^r \prod_{j=1}^{k} q_j \right\}^{1/2}}{\Gamma\left(\frac{k+1}{2}\right)},$$

where $N(\eta)$ is the number of members in $\{u : u \in L,\ u \in G_\eta^\phi\}$, and Γ is the gamma function.

The expansion (2.6) and $\hat{K}_2$ given by (2.7) were derived by Menéndez et al. (1997). However, the necessary conditions for Theorem 2.2 about the function ϕ were not shown in their paper. Although the calculus of K_3 in the case of $G = G_\eta^\phi$ is too complicated, we know by Theorem 2.1 that $K_3 = O(n^{-1})$. Therefore, we obtain

$$K_1' + \hat{K}_2, \tag{2.8}$$

where

$$K_1' = \Pr\{\chi_r^2 < \eta\} + \frac{1}{24}n^{-1}\sum_{j=0}^{3} r_j' \Pr\{\chi_{r+2j}^2 < \eta\},$$

as an asymptotic approximation to $\Pr\{C_\phi(U) < \eta | H_0\}$. The approximation $\hat{K}_2$ does not look so complicated. However, it is very hard to calculate the value of $\hat{K}_2$ when n and k are not small, because of the difficulty in counting $N(\eta)$ in (2.7). Thus, the use of the approximation is not so practical, except when n and k are small.

The numerical accuracy of the approximation was shown by Yarnold (1972) for Pearson's X^2 statistic and by Read^b (1984) and Read and Cressie (1988) for power divergence statistics.

Yarnold (1972) numerically examined the accuracy of (a) the chi-square approximation, (b) an Edgeworth approximation for the null distribution of Pearson's X^2, and (c) an approximation based on Edgeworth expansion and discontinuous terms. He proposed the use of (c).

From the numerical results obtained by Yarnold (1972), we notice that the chi-square approximation rarely performs better than the Edgeworth approximation. Thus, an Edgeworth approximation appears to be an effective approximation of the null distribution of the above statistics when the discontinuous term in the asymptotic expansion cannot be expressed in a simple form. When it can, Yarnold's recommendation applies. It is very difficult to represent the discontinuous term in a simple form for the earlier mentioned statistics under alternative hypotheses and also for more general multinomial models such as contingency tables under a null hypothesis. A mathematical explanation under alternative hypotheses is provided in Taneichi et al. (1984) and a mathematical explanation for a 2-way contingency table will be provided in Chap. 3. Edgeworth approximations of the distributions of specific multinomial goodness-of-fit statistics under alternative hypotheses were investigated (Taneichi et al. 2001, 2002; Sekiya and Taneichi 2004; Pan et al. 2011). Based on numerical investigations, it was found that omission of the discontinuous term does not lead to a serious error.

2.6 Transformed Statistics Based on an Asymptotic Expansion

In order to improve small-sample accuracy of the chi-square approximation of the distribution of a test statistic, a Bartlett-type transformation and a transformation called improved transformation were considered.

We consider a non-negative random variable T that has an asymptotic expansion

$$\Pr\{T \le z\} = \Pr\{\chi_f^2 \le z\} + n^{-1} \sum_{\omega=0}^{\zeta} a_\omega \Pr\{\chi_{f+2\omega}^2 \le z\} + O(n^{-2}), \tag{2.9}$$

where ζ is a positive integer, and $a_0, \dots, a_\zeta$ do not depend on n (> 0) and satisfy $a_0 + \cdots + a_\zeta = 0$.

When $\zeta = 1$ in (2.9), the Bartlett adjustment T_B (Lawley 1956; Barndorff-Nielsen and Cox 1984; Barndorff-Nielsen and Hall 1988) is defined as

$$T_B(T; a_0, f, n) = \left(1 + \frac{2a_0}{nf}\right) T. \tag{2.10}$$

When $\zeta = 3$ in (2.9), improved transformation T_I (Kakizawa 1996; Fujikoshi 1997, 2000) and Bartlett-type adjustment T_{CF} (Cordeiro and Ferrari 1991) are defined as

$$\begin{aligned} T_I(T; a_0, a_2, a_3, f, n) =& (n\alpha + \beta)^2 \ln\left[1 + \frac{1}{(n\alpha)^2}\left\{T + \frac{1}{n\alpha}(T^2 + \gamma T^3)\right.\right. \\ &\left.\left. + \frac{1}{(n\alpha)^2}\left(\frac{1}{3}T^3 + \frac{3\gamma}{4}T^4 + \frac{9\gamma^2}{20}T^5\right)\right\}\right], \end{aligned} \tag{2.11}$$

where $\alpha = -f(f+2)/\{2(a_2 + a_3)\}$, $\beta = -(f+2)a_0/\{2(a_2 + a_3)\}$ and $\gamma = a_3/\{(f+4)(a_2 + a_3)\}$, and

$$T_{CF}(T; a_0, a_1, a_2, f, n) = \left\{1 + \frac{2a_0}{nf} + \frac{2(a_0 + a_1)}{nf(f+2)}T + \frac{2(a_0 + a_1 + a_2)}{nf(f+2)(f+4)}T^2\right\} T, \tag{2.12}$$

respectively. Then, for all $\xi \in \{B, I, CF\}$,

$$\Pr\{T_\xi \le z\} = \Pr\{\chi_f^2 \le z\} + O(n^{-2}), \tag{2.13}$$

where χ_f^2 denotes a chi-square random variable with degrees of freedom f.

References

Ali SM, Silvey SD (1966) A general class of coefficients of divergence of one distribution from another. J R Stat Soc Ser B 28:131–142

Barndorff-Nielsen OE, Cox DR (1984) Bartlett adjustments to the likelihood ratio statistic and the distribution of maximum likelihood estimator. J R Stat Soc Ser B 46:483–495

Barndorff-Nielsen OE, Hall P (1988) On the level-error after Bartlett adjustment of the likelihood ratio statistic. Biometrika 75:374–378

Bhattacharya RN, Ranga Rao R (1976) Normal approximation and asymptotic expansions. Wiley, New York

Birch MW (1964) A new proof of the Pearson-Fisher theorem. Ann Math Stat 35:817–824

Cordeiro GM, Ferrari SLP (1991) A modified score test statistic having chi-squared distribution to order n^{-1}. Biometrika 78:573–582

Cressie N, Read TRC (1984) Multinomial goodness-of-fit tests. J Roy Statist Soc Ser B 46:440–464

Csiszár I (1967) Information-type measures of difference of probability distributions and indirect observations. Stud Sci Math Hung 2:299–318

Esséen CG (1945) Fourier analysis of distribution functions: a mathematical study of the Laplace-Gaussian law. Acta Math 77:1–125

Fujikoshi Y (1997) A method for improving the large-sample chi-squared approximations to some multivariate test statistics. Amer J Math Management Sci 17:15–29

Fujikoshi Y (2000) Transformations with improved chi-squared approximations. J Multivariate Anal 72:249–263

Kakizawa Y (1996) Higher order monotone Bartlett-type adjustment for some multivariate test statistics. Biometrika 83(4):923–927

Lawley DN (1956) A general method for approximating to the distribution of the likelihood ratio criteria. Biometrika 43:295–303

Lin J (1991) Divergence measures based on the Shannon entropy. IEEE Trans Inf Theory 37:145–151

Menéndez ML, Pardo JA, Pardo L, Pardo MC (1997) Asymptotic approximations for the distributions of the (h, ϕ)-divergence goodness-of-fit statistics: application to Renyi's statistic. Kybernetes 26:442–452

Pan ei htwe, Taneichi N, Sekiya Y, (2011) Improved approximation for the distribution of multinomial goodness-of-fit statistics based on ϕ-divergence under nonlocal alternatives. J Jpn Stat Soc 41(2):121–142

Pardo L, Pardo MC, Zografos K (1999) Homogeneity for multinomial populations based on ϕ-divergences. J Jpn Stat Soc 29:213–228

Pardo L (2006) Statistical inference based on divergence measures. Chapman & Hall/CRC

Ranga Rao R (1961) On the central limit theorem in R_k. Bull Amer Math Soc 67:359–361

Read TRC (1984) Closer asymptotic approximations for the distributions of the power divergence goodness-of-fit statistics. Ann Inst Statist Math 36:59–69

Read TRC (1984) Small-sample comparisons for the power divergence goodness-of-fit statistics. J Am Stat Assoc 79:929–935

Read TRC, Cressie NAC (1988) Goodness-of-fit statistics for discrete multivariate data. Springer, New York

Rukhin AL (1994) Optimal estimator for the mixture parameter by the method of moments and information affinity. In: Trans. 12th Prague conference on information theory, pp. 214–219

Sekiya Y, Taneichi N (2004) Improvement of approximations for the distributions of multinomial goodness-of-fit statistics under nonlocal alternatives. J Multivariate Anal 91:199–223

Siotani M, Fujikoshi Y (1984) Asymptotic approximations for the distributions of multinomial goodness-of-fit statistics. Hiroshima Math J 14:115–124

Taneichi N, Sekiya Y, Suzukawa A (2001) An asymptotic approximation for the distribution of ϕ-divergence multinomial goodness-of-fit statistic under local alternatives. J Jpn Stat Soc 31(2):207–224

Taneichi N, Sekiya Y, Suzukawa A (2002) Asymptotic approximations for the distributions of the multinomial goodness-of-fit statistics under local alternatives. J Multivariate Anal 81:335–359

Yarnold JK (1972) Asymptotic approximations for the probability that a sum of lattice random vectors lies in a convex set. Ann Math Stat 43:1566–1580

Zografos K, Ferentions K, Papaioannou T (1990) Sampling properties and multinomial goodness-of-fit and divergence tests. Commun Stat-Theory Methods 19:1785–1802

Chapter 3
A Transformed Statistic for the Test of Independence in a $J \times K$ Contingency Table

Keywords Asymptotic expansion · Local edgeworth approximation · Test of independence · Transformed statistics · 2-way contingency table

3.1 Statistic for the Test of Independence in a 2-Way Contingency Table Based on ϕ-Divergence

In a generic $J \times K$ contingency table (Table 3.1), let $X = (X_{11}, \ldots, X_{1K}, \ldots, X_{J1}, \ldots, X_{JK})^\top$ be distributed according to a multinomial distribution $\mathcal{M}_{JK}$ $(n; p_{11}, \ldots, p_{1K}, \ldots, p_{J1}, \ldots, p_{JK})$, where $\sum_{j=1}^{J}\sum_{k=1}^{K} X_{jk} = n$, $0 < p_{jk} < 1$, $j = 1, \ldots, J$, $k = 1, \ldots, K$, and $\sum_{j=1}^{J}\sum_{k=1}^{K} p_{jk} = 1$, and let the marginal probabilities

Table 3.1 A generic $J \times K$ contingency table

						Total
	X_{11}	$\cdots$	X_{1k}	$\cdots$	X_{1K}	$X_{1\cdot}$
	$\vdots$	$\vdots$	$\vdots$	$\vdots$	$\vdots$	$\vdots$
	X_{j1}	$\cdots$	X_{jk}	$\cdots$	X_{jK}	$X_{j\cdot}$
	$\vdots$	$\vdots$	$\vdots$	$\vdots$	$\vdots$	$\vdots$
	X_{J1}	$\cdots$	X_{Jk}	$\cdots$	X_{JK}	$X_{J\cdot}$
Total	$X_{\cdot 1}$	$\cdots$	$X_{\cdot k}$	$\cdots$	$X_{\cdot K}$	n

This chapter is based on the following article: Journal of Multivariate Analysis, 98, Taneichi N. and Sekiya Y., Improved transformed statistics for the test of independence in $r \times s$ contingency tables, 1630–1657, Copyright Elsevier (2007).

N. Taneichi and Y. Sekiya, *Improving Tests for Discrete Small Sample Data*,
JSS Research Series in Statistics,
https://doi.org/10.1007/978-981-95-5301-3_3

of rows and columns be $p_{j\cdot} = \sum_{k=1}^{K} p_{jk}, j = 1, \ldots, J$ and $p_{\cdot k} = \sum_{j=1}^{J} p_{jk}, k = 1, \ldots, K$, respectively. Then the null hypothesis of independence is

$$H_0^{(I)} : p_{jk} = p_{j\cdot} p_{\cdot k}, \quad j = 1, \ldots, J, \ k = 1, \ldots, K. \tag{3.1}$$

For testing the hypothesis $H_0^{(I)}$, we consider a class of statistics $C_\phi^{[2]}$ based on ϕ-divergence. We denote the unrestricted maximum likelihood estimators of p_{jk} by $\hat{p}_{jk}$ and the maximum likelihood estimators of $p_{j\cdot}$ and $p_{\cdot k}$ under $H_0^{(I)}$ by $\hat{p}_{j\cdot}$ and $\hat{p}_{\cdot k}$, respectively, i.e., $\hat{p}_{jk} = X_{jk}/n, j = 1, \ldots, J, k = 1, \ldots, K$, $\hat{p}_{j\cdot} = X_{j\cdot}/n, j = 1, \ldots, J$, and $\hat{p}_{\cdot k} = X_{\cdot k}/n, k = 1, \ldots, K$, where $X_{j\cdot} = \sum_{k=1}^{K} X_{jk}$ and $X_{\cdot k} = \sum_{j=1}^{J} X_{jk}$. Then the ϕ-divergence statistics $C_\phi^{[2]}$ are given by

$$C_\phi^{[2]} = 2n \sum_{j=1}^{J} \sum_{k=1}^{K} \hat{p}_{j\cdot} \hat{p}_{\cdot k} \phi \left(\frac{\hat{p}_{jk}}{\hat{p}_{j\cdot} \hat{p}_{\cdot k}} \right), \tag{3.2}$$

Zografos (1993). When we choose a convex function ϕ^a given by (2.3) as ϕ, $C_{\phi^a}^{[2]}$ become statistics based on power divergence as follows (Read and Cressie 1988, pp. 23–24).

$$R^{a[2]} \equiv C_{\phi^a}^{[2]} = 2n \sum_{j=1}^{J} \sum_{k=1}^{K} I^a(\hat{p}_{jk}, \hat{p}_{j\cdot} \hat{p}_{\cdot k}), \tag{3.3}$$

where

$$I^a(e, f) = \begin{cases} \{a(a+1)\}^{-1} e \left\{ \left(\frac{e}{f} \right)^a - 1 \right\} & (a \neq 0, -1) \\ e \ln \left(\frac{e}{f} \right) & (a = 0) \\ f \ln \left(\frac{f}{e} \right) & (a = -1). \end{cases} \tag{3.4}$$

It is immediately shown that $R^{0[2]}$ is the log likelihood ratio statistic and that $R^{1[2]}$ is Pearson's X^2 statistic. Statistic $R^{2/3[2]}$ corresponds to the statistic recommended by Cressie and Read (1984) for the goodness-of-fit test. Under the null hypothesis $H_0^{(I)}$ given by (3.1), it is known that all members of the class of statistics $C_\phi^{[2]}$ defined by (3.2) are asymptotically distributed as a chi-square distribution with degrees of freedom $(J-1)(K-1)$.

3.2 Local Edgeworth Approximation

We consider a local Edgeworth approximation for the probability of $X = (X_{11}, \ldots, X_{1K}, \ldots, X_{J1}, \ldots, X_{JK})^\top$ under the null hypothesis $H_0^{(I)}$. Let X be distributed according to $\mathcal{M}_{JK}(n; p_{1\cdot} p_{\cdot 1}, \ldots, p_{1\cdot} p_{\cdot K}, \ldots, p_{J\cdot} p_{\cdot 1}, \ldots, p_{J\cdot} p_{\cdot K})$.

Let

$$U_{jk} = n^{-1/2}(X_{jk} - np_{j\cdot}p_{\cdot k}), \quad j = 1, \ldots, J, \ k = 1, \ldots, K. \tag{3.5}$$

Then $U = (U_{11}, \ldots, U_{1K}, \ldots, U_{J1}, \ldots, U_{J,K-1})^\top$ is a lattice random vector that takes values in the set

$$L = \left\{u = (u_{11}, \ldots, u_{1K}, \ldots, u_{J1}, \ldots, u_{J,K-1})^\top : u = n^{-1/2}(\tilde{x} - n\tilde{p}), \ \tilde{x} \in M\right\},$$

where

$$\tilde{p} = (p_{1\cdot}p_{\cdot 1}, \ldots, p_{1\cdot}p_{\cdot K}, \ldots, p_{J\cdot}p_{\cdot 1}, \ldots, p_{J\cdot}p_{\cdot K-1})^\top \tag{3.6}$$

and

$$M = \Bigg\{\tilde{x} = (x_{11}, \ldots, x_{1K}, \ldots, x_{J1}, \ldots, x_{J,K-1})^\top : x_{jk}\text{'s are non-negative integers and } x_{11} + \cdots + x_{1K} + \cdots + x_{J1} + \cdots + x_{J,K-1} \le n\Bigg\}.$$

The following theorem can be obtained from Lemma 2.1 of Siotani and Fujikoshi (1984) by changing $p_0 = (p_{01}, \cdots, p_{0k-1})$ to $\tilde{p}$.

Theorem 3.1 *For each $\tilde{x} \in M$, let $u = n^{-1/2}(\tilde{x} - n\tilde{p})$. Then*

$$\Pr\{U = u | H_0^{(I)}\} = n^{-(JK-1)/2} f(u) \left\{1 + n^{-1/2}h_1(u) + n^{-1}h_2(u) + o(n^{-1})\right\},$$

where

$$f(u) = (2\pi)^{-(JK-1)/2}|\Omega|^{-1/2} \exp\left(-\frac{1}{2}u^\top \Omega^{-1} u\right), \tag{3.7}$$

$$\Omega = \mathrm{diag}(p_{1\cdot}p_{\cdot 1}, \ldots, p_{1\cdot}p_{\cdot K}, \ldots, p_{J\cdot}p_{\cdot 1}, \ldots, p_{J\cdot}p_{\cdot K-1}) - \tilde{p}\tilde{p}^\top, \tag{3.8}$$

$$h_1(u) = -\frac{1}{2}\sum_{j=1}^{J}\sum_{k=1}^{K}\frac{u_{jk}}{p_{j\cdot}p_{\cdot k}} + \frac{1}{6}\sum_{j=1}^{J}\sum_{k=1}^{K}\frac{u_{jk}^3}{p_{j\cdot}^2 p_{\cdot k}^2}, \tag{3.9}$$

$$h_2(u) = \frac{1}{2}\{h_1(u)\}^2 + \frac{1}{12}(1 - q_1^{[2]}q_2^{[2]}) + \frac{1}{4}\sum_{j=1}^{J}\sum_{k=1}^{K}\frac{u_{jk}^2}{p_{j\cdot}^2 p_{\cdot k}^2} - \frac{1}{12}\sum_{j=1}^{J}\sum_{k=1}^{K}\frac{u_{jk}^4}{p_{j\cdot}^3 p_{\cdot k}^3}, \tag{3.10}$$

$$u_{JK} = -(u_{11} + \cdots + u_{1K} + \cdots + u_{J1} + \cdots + u_{J,K-1}), \tag{3.11}$$

$$q_1^{[2]} = \sum_{j=1}^{J} p_{j\cdot}^{-1}, \tag{3.12}$$

$$q_2^{[2]} = \sum_{k=1}^{K} p_{\cdot k}^{-1}, \tag{3.13}$$

and $\tilde{p}$ is given by (3.6).

3.3 An Approximation for the Distribution of $C_\phi^{[2]}$ Under $H_0^{(I)}$ Based on an Asymptotic Expansion

We derive an approximation based on an asymptotic expansion for the distribution of $C_\phi^{[2]}$ under $H_0^{(I)}$. We consider the following approximation for the distribution of $C_\phi^{[2]}$ under $H_0^{(I)}$ corresponding to approximations (2.5) for the goodness-of-fit test.

$$\Pr\{C_\phi^{[2]} \le b | H_0\} \approx J_1^* + J_2^*,$$

where the J_1^* term is a multivariate Edgeworth expansion assuming a continuous distribution and the J_2^* term, which corresponds to the K_2 term given by (2.5) in the case of a goodness-of-fit test, is a discontinuous term to account for the discontinuity. With regard to evaluation of the J_1^* term, we obtain the following theorem.

Theorem 3.2 *If ϕ is four times continuously differentiable, the J_1^* term is evaluated as*

$$J_1^* = \Pr\{\chi^2_{(J-1)(K-1)} \le b\} + n^{-1} \sum_{\ell=0}^{3} w_\ell^\phi \Pr\{\chi^2_{(J-1)(K-1)+2\ell} \le b\} + o(n^{-1}), \tag{3.14}$$

where

$$\begin{aligned}
w_0^\phi &= -\tfrac{1}{12}\{(V-1)(W-1)\},\\
w_1^\phi &= \tfrac{1}{24}\big[3\{(V-J^2)(W-K^2) + 2(V-J)(W-K)\}\\
&\quad +6\phi'''(1)(V-J^2)(W-K^2)\\
&\quad +\{\phi'''(1)\}^2\{3(V-J^2)(W-K^2) + 2(V-3J+2)(W-3K+2)\}\\
&\quad -3\phi^{(4)}(1)(V-2J+1)(W-2K+1)\big],\\
w_2^\phi &= \tfrac{1}{24}\big[-6\{(V-J^2)(W-K^2) + (V-2J+1)(W-2K+1)\}\\
&\quad -4\phi'''(1)\{3(V-J^2)(W-K^2) + (V-3J+2)(W-3K+2)\}\\
&\quad -2\{\phi'''(1)\}^2\{3(V-J^2)(W-K^2) + 2(V-3J+2)(W-3K+2)\}\\
&\quad +3\phi^{(4)}(1)(V-2J+1)(W-2K+1)\big],\\
w_3^\phi &= \tfrac{1}{24}\big[\{1+\phi'''(1)\}^2\{3(V-J^2)(W-K^2) + 2(V-3J+2)(W-3K+2)\}\big],
\end{aligned}$$

and V being $q_1^{[2]}$ given by (3.12) and W being $q_2^{[2]}$ given by (3.13).

Proof of Theorem 3.2 is shown in Appendix **??**.

If we apply ϕ^a as ϕ in Theorem 3.2, we obtain the following corollary for the statistics based on power divergence.

Corollary 3.1 *When the statistic is $R^{a[2]}$ given by (3.3), the J_1^* term is evaluated as*

$$J_1^* = \Pr\{\chi^2_{(J-1)(K-1)} \le b\} + n^{-1}\sum_{\ell=0}^{3} w_\ell^{(a)} \Pr\{\chi^2_{(J-1)(K-1)+2\ell} \le b\} + o(n^{-1}),$$

where $w_\ell^{(a)}$, $\ell = 0, 1, 2, 3$ are defined as w_ℓ^ϕ, $\ell = 0, 1, 2, 3$ in the case of $\phi'''(1) = a - 1$ and $\phi^{(4)}(1) = (a-1)(a-2)$, respectively.

For $b > 0$, let $B_\phi(b)$ be a set defined by

$$B_\phi(b) = \{u^* = (u_1^*, \ldots, u_{JK-1}^*)^\top : C_\phi^{[2]}(u) \equiv C_\phi^{[2]}(u^*) < b\}, \tag{3.15}$$

where

$$u_\ell^* = u_{jk},\ \ell = (j-1)K + k \text{ for } j = 1, \ldots, J,\ k = 1, \ldots, K, \tag{3.16}$$

and $C_\phi^{[2]}(u)$ is given by substituting u for U in $C_\phi^{[2]}(U)$ defined in (A.1). Consider the set $B_\ell \subset R_{JK-2}$, ($\ell = 1, \ldots, JK-1$) and continuous functions η_ℓ and θ_ℓ on R_{JK-2} into R_1 such that $B_\phi(b)$ defined by (3.15) is represented as

$$\begin{aligned} B_\phi(b) = \{u^* = (u_1^*, \ldots, u_{JK-1}^*)^\top : \eta_\ell(\tilde{u}_\ell^*) < u_\ell^* < \theta_\ell(\tilde{u}_\ell^*),\\ \tilde{u}_\ell^* = (u_1^*, \ldots, u_{\ell-1}^*, u_{\ell+1}^*, \ldots, u_{JK-1}^*)^\top \in B_\ell\}. \end{aligned} \tag{3.17}$$

Then,

$$\begin{aligned} J_2^* = -n^{-1/2}\sum_{\ell=1}^{JK-1} n^{-(JK-1-\ell)/2} \sum_{u_{\ell+1}^* \in M_{\ell+1}^\phi} \cdots \sum_{u_{JK-1}^* \in M_{JK-1}^\phi} \\ \times \int \cdots \int_{B_\ell} [S_1(\sqrt{n}u_\ell^* + np_\ell^*) f(u^*)]_{\eta_\ell(\tilde{u}_\ell^*)}^{\theta_\ell(\tilde{u}_\ell^*)} du_1^* \cdots du_{\ell-1}^*, \end{aligned} \tag{3.18}$$

where

$$M_\ell^\phi = B_\phi(b) \cap L_\ell,\quad \ell = 1, \ldots, JK-1,$$

$$L_\ell = \left\{y_\ell : y_\ell = n^{-1/2}(n_\ell - np_\ell^*),\ n_\ell \text{ is an integer}\right\}, \tag{3.19}$$

$$p_\ell^* = p_{j\cdot}p_{\cdot k},\ \ell = (j-1)K + k \text{ for } j = 1, \ldots, J,\ k = 1, \ldots, K, \tag{3.20}$$

$$S_1(t) = t - [t] - \frac{1}{2}, \tag{3.21}$$

and f being defined by (3.7). In order to evaluate the J_2^* term of the test statistics using the same method as that of Yarnold (1972), it is necessary to show

$$[S_1(\sqrt{n}u_\ell^* + np_\ell^*) f(u^*)]_{\eta_\ell(\tilde{u}_\ell^*)}^{\theta_\ell(\tilde{u}_\ell^*)} = c[S_1(\sqrt{n}u_\ell^* + np_\ell^*)]_{\eta_\ell(\tilde{u}_\ell^*)}^{\theta_\ell(\tilde{u}_\ell^*)} + o(1), \tag{3.22}$$

where c is a constant. From (A.2), (A.3) and the fact that $\eta_\ell(\tilde{u}_\ell^*)$ and $\theta_\ell(\tilde{u}_\ell^*)$ in (3.17) are values of u_ℓ^* such that $C_\phi^{[2]}(u) = C_\phi^{[2]}(u^*) = b$, where b is a constant, we find that

$$u^\top \Omega^{-1} u = b + \sum_{j=1}^{J} \frac{u_{j\cdot}^2}{p_{j\cdot}} + \sum_{k=1}^{K} \frac{u_{\cdot k}^2}{p_{\cdot k}} + o(1)$$

when $u_\ell^* = \eta_\ell(\tilde{u}_\ell^*)$ or $u_\ell^* = \theta_\ell(\tilde{u}_\ell^*)$. Therefore, we can not evaluate $u^\top \Omega^{-1} u$ as a constant when $u_\ell^* = \eta_\ell(\tilde{u}_\ell^*)$ or $u_\ell^* = \theta_\ell(\tilde{u}_\ell^*)$. Since $f(u)$ is a constant if and only if $u^\top \Omega^{-1} u$ is a constant, we can not prove (3.22). Thus, unlike the null distribution of goodness-of-fit test statistics, we can not obtain a simple form of approximation of J_2^* such as $\hat{K}_2$ given by (2.7).

By another method of Yarnold (1972), J_2^* is evaluated as follows.

Theorem 3.3 *The J_2^* term can be represented in the following form.*

$$J_2^* = A - B + C + O\left(n^{-1}\right), \tag{3.23}$$

where

$$A = n^{-(JK-1)/2} \sum_{u_1^* \in M_1^\phi} \cdots \sum_{u_{JK-1}^* \in M_{JK-1}^\phi} f(u^*),$$

$$B = \Pr\{\chi^2_{(J-1)(K-1)} \leq b\},$$

and

$$C = n^{-1/2} \int \cdots \int_{B_\phi(b)} S_1(\sqrt{n}u_{JK-1}^* + np_{JK-1}^*) \left(\frac{u_{JK-1}^*}{p_{JK-1}^*} - \frac{u_{JK}^*}{p_{JK}^*} \right) f(u^*) du^*, \tag{3.24}$$

where $du^* = du_1^* \cdots du_{JK-1}^*$.

Proof of Theorem 3.3 is shown in Appendix **??**.

With regard to the C term in Theorem 3.3, we obtain the following theorem.

Theorem 3.4 *Let*

$$\rho(j,k) = \left(\frac{p_{j\cdot}}{1 - p_{j\cdot}} + \frac{p_{\cdot k}}{1 - p_{\cdot k}} \right)^{-1}.$$

If

$$\rho(j,k) < 1 \ \textit{for}\ j = 1, \ldots, J,\ k = 1, \ldots, K, \tag{3.25}$$

then the C term in Theorem 3.3 satisfies the following,

$$|C| \leq \Delta + O\left(n^{-1}\right), \tag{3.26}$$

where

$$\Delta = n^{-1/2} \frac{1}{\sqrt{2\pi}} \sum_{\ell=0}^{\infty} \Pr\{\chi^2_{(J-1)(K-1)+2\ell} \leq b\}$$
$$\times \left\{ \frac{1}{p_{J\cdot} p_{\cdot K}} \sum_{j=1}^{J} \sum_{k=1}^{K} \sqrt{p_{j\cdot} p_{\cdot k} (p_{j\cdot} + p_{\cdot k} - 2 p_{j\cdot} p_{\cdot k})}\, \tau_{(j,k)}^{(\ell)} \right.$$
$$\left. - \sqrt{\frac{p_{J\cdot} + p_{\cdot K} - 2 p_{J\cdot} p_{\cdot K}}{p_{J\cdot} p_{\cdot K}}}\, \tau_{(J,K)}^{(\ell)} + \sqrt{\frac{p_{J\cdot} + p_{\cdot K-1} - 2 p_{J\cdot} p_{\cdot K-1}}{p_{J\cdot} p_{\cdot K-1}}}\, \tau_{(J,K-1)}^{(\ell)} \right\},$$

$$\tau_{(j,k)}^{(\ell)} = \{\rho(j,k)\}^{\ell} \binom{1/2}{\ell},$$

and $\binom{a}{b}$ *are the binomial coefficients.*

Proof of Theorem 3.4 is shown in Appendix **??**.

According to Theorem 3.3 and Theorem 3.4, the J_2^* term is approximated well by $A - B$ when condition (3.25) is satisfied and Δ is sufficiently small. However, condition (3.25) is never satisfied in $J \times K$ contingency tables when $J \geq 3$ and $K \geq 3$. Furthermore, in 2×2 or 2×3 or 3×2 contingency tables, condition (3.25) is satisfied in only restricted domains of $p_{j\cdot}$ and $p_{\cdot k}$. Therefore, for the test of independence in contingency tables, it is very difficult to obtain an appropriate approximation for the J_2^* term.

Regarding the goodness-of-fit test for a multinomial distribution, Yarnold (1972) numerically examined the accuracy of approximations given by $K_1' + \hat{K}_2$ of (2.8), chi-square approximation, and Edgeworth approximation assuming a continuous distribution K_1' of (2.8) and concluded that the approximation given by $K_1' + \hat{K}_2$ performed better than the others in the case of Pearson's X^2 statistic. However, as shown above, for the test of independence in contingency tables, it is very difficult to obtain an appropriate approximation such as $\hat{K}_2$ in the case of the goodness-of-fit test for a multinomial distribution, and, besides, chi-square approximation rarely performs better than Edgeworth approximation assuming a continuous distribution in the numerical results of Yarnold (1972). Furthermore, Taneichi et al. (2001) and Sekiya and Taneichi (2004) numerically showed that Edgeworth approximation assuming a continuous distribution performs better than does non-central chi-square approximation for the multinomial goodness-of-fit test under local and non local alternatives. Also, Taneichi and Sekiya (2007) numerically showed that for the test of independence in a 2-way contingency table, omission of the discontinuous term does not lead to a serious error.

Consequently, on the basis of the numerical results of the multinomial goodness-of-fit test, we propose the use of J_1^* as an approximation for the distribution of $C_\phi^{[2]}$ under $H_0^{(I)}$, that is,

$$\Pr\{C_\phi^{[2]} \le b | H_0^{(I)}\} \approx J_1^*. \tag{3.27}$$

3.4 A Transformed Statistic Based on the Approximation

In this section, applying (2.10), (2.11), and (2.12) in Chap. 2 to the evaluation (3.14) given by Theorem 3.2, we construct transformations for improving small-sample accuracy of the chi-square approximation of the distribution of $C_\phi^{[2]}$ under $H_0^{(I)}$.

Since equation $\sum_{\ell=0}^{3} w_\ell^\phi = 0$ holds in Theorem 3.2, we obtain the following transformed statistics $C_\phi^{I[2]}$ and $C_\phi^{CF[2]}$ from (2.11) and (2.12), respectively.

$$C_\phi^{I[2]} = (n\alpha + \beta)^2 \ln(1 + Q), \tag{3.28}$$

where

$$Q = \frac{1}{(n\alpha)^2}\Bigg[C_\phi^{[2]} + \frac{1}{n\alpha}\left\{(C_\phi^{[2]})^2 + \gamma (C_\phi^{[2]})^3\right\} + \frac{1}{(n\alpha)^2}\left\{\frac{1}{3}(C_\phi^{[2]})^3 + \frac{3\gamma}{4}(C_\phi^{[2]})^4 + \frac{9\gamma^2}{20}(C_\phi^{[2]})^5\right\}\Bigg],$$

$$\alpha = -\frac{D(D+2)}{2(w_2^\phi + w_3^\phi)},\quad \beta = -\frac{(D+2)w_0^\phi}{2(w_2^\phi + w_3^\phi)},\quad \text{and}\quad \gamma = \frac{w_3^\phi}{(D+4)(w_2^\phi + w_3^\phi)},$$

and

$$C_\phi^{CF[2]} = \left\{1 + \frac{2w_0^\phi}{nD} + \frac{2(w_0^\phi + w_1^\phi)}{nD(D+2)}C_\phi^{[2]} + \frac{2(w_0^\phi + w_1^\phi + w_2^\phi)}{nD(D+2)(D+4)}(C_\phi^{[2]})^2\right\}C_\phi^{[2]}, \tag{3.29}$$

where $D = (J-1)(K-1)$. Therefore, we propose the following approximations.

$$\Pr\{C_\phi^{I[2]} \le b | H_0^{(I)}\} \approx \Pr\{\chi^2_{(J-1)(K-1)} \le b\},$$

and

$$\Pr\{C_\phi^{CF[2]} \le b | H_0^{(I)}\} \approx \Pr\{\chi^2_{(J-1)(K-1)} \le b\}.$$

When statistics are $R^{a[2]}$ $(a \ne 0)$ given by (3.3), by substituting $w_\ell^{(a)}$, $\ell = 0, 1, 2, 3$ for w_ℓ^ϕ, $\ell = 0, 1, 2, 3$ and $R^{a[2]}$ for $C_\phi^{[2]}$ in (3.28) and (3.29), we obtain transformed statistics $R_I^{a[2]}$ and $R_{CF}^{a[2]}$ as a special case of $C_\phi^{I[2]}$ and $C_\phi^{CF[2]}$, respectively. When the statistic is $R^{0[2]}$ (the log likelihood ratio statistic), equations $w_2^{(0)} = w_3^{(0)} = 0$ and $w_0^{(0)} + w_1^{(0)} = 0$ hold in Corollary 3.1. Therefore, we can adopt Bartlett adjustment as a transformed statistic and we obtain the following statistic from (2.10).

$$R_B^{0[2]} = \left\{1 + \frac{2w_0^{(0)}}{n(J-1)(K-1)}\right\} R^{0[2]}.$$

As well as $R_I^{a[2]}$ and $R_{CF}^{a[2]}$ $(a \neq 0)$, we propose the following approximation for the statistic $R_B^{0[2]}$.

$$\Pr\{R_B^{0[2]} \leq b | H_0^{(I)}\} \approx \Pr\{\chi^2_{(J-1)(K-1)} \leq b\}.$$

We also note that $R_{CF}^{0[2]} = R_B^{0[2]}$, then $R_{CF}^{a[2]}$ can be defined when $a = 0$ as well as when $a \neq 0$.

Now, w_ℓ^ϕ, $\ell = 0, 1, 2, 3$ include $V \equiv q_1^{[2]}$ and $W \equiv q_2^{[2]}$, which are functions of unknown parameters $p_{j\cdot}$, $j = 1, \ldots, J$ and $p_{\cdot k}$, $k = 1, \ldots, K$. Then, in practical applications, we must estimate V and W. By substituting $\hat{p}_{j\cdot} = X_{j\cdot}/n$ and $\hat{p}_{\cdot k} = X_{\cdot k}/n$ for $p_{j\cdot}$ and $p_{\cdot k}$, respectively, we use the maximum likelihood estimators of (3.12) and (3.13), that is,

$$\hat{V} = \sum_{j=1}^{J} \hat{p}_{j\cdot}^{-1}$$

and

$$\hat{W} = \sum_{k=1}^{K} \hat{p}_{\cdot k}^{-1}.$$

In Taneichi and Sekiya (2007), the small-sample performance of approximation by a chi-square distribution of a test statistic for a test of independence of a 4×4 contingency table were investigated. For power divergence statistics, performance of transformed statistics $R_I^{a[2]}$ $(a = 0, 2/3, 1)$ were numerically compared with that of the original statistics $R^{a[2]}$ $(a = 0, 2/3, 1)$ for significance levels 0.01 and 0.05. The result of comparisons was as follows. Transformed statistics $R_I^{a[2]}$ $(a = 0, 2/3, 1)$ performed much better than the original statistics $R^{a[2]}$ $(a = 0, 2/3, 1)$. Furthermore, in Taneichi and Sekiya (2007), power against some alternative hypotheses of transformed statistics $R_I^{a[2]}$ $(a = 0, 2/3, 1)$ were compared with those of the original statistics $R^{a[2]}$ $(a = 0, 2/3, 1)$. The result of comparisons was as follows. The power of the transformed statistics $R_I^{a[2]}$ $(a = 0, 2/3, 1)$ were almost the same as that of the original statistics $R^{a[2]}$ $(a = 0, 2/3, 1)$.

References

Cressie N, Read TRC (1984) Multinomial goodness-of-fit tests. J Roy Statist Soc Ser B 46:440–464

Read TRC, Cressie NAC (1988) Goodness-of-fit statistics for discrete multivariate data. Springer, New York

Sekiya Y, Taneichi N (2004) Improvement of approximations for the distributions of multinomial goodness-of-fit statistics under nonlocal alternatives. J Multivariate Anal 91:199–223

Siotani M, Fujikoshi Y (1984) Asymptotic approximations for the distributions of multinomial goodness-of-fit statistics. Hiroshima Math J 14:115–124

Taneichi N, Sekiya Y, Suzukawa A (2001) An asymptotic approximation for the distribution of ϕ-divergence multinomial goodness-of-fit statistic under local alternatives. J Jpn Stat Soc 31(2):207–224

Taneichi N, Sekiya Y (2007) Improved transformed statistics for the test of independence in $r \times s$ contingency tables. J Multivariate Anal 98:1630–1657

Yarnold JK (1972) Asymptotic approximations for the probability that a sum of lattice random vectors lies in a convex set. Ann Math Stat 43:1566–1580

Zografos K (1993) Asymptotic properties of Φ-divergence statistic and its application in contingency tables. Internat J Math Statist Sci 2:5–21

Chapter 4
A Transformed Statistic for the Test of Complete Independence in a Contingency Table

Keywords Asymptotic expansion · Complete independence · Independence among groups of factors · Local Edgeworth expansion · M-way contingency table

4.1 Statistic for the Test of Two Kinds of Independence in a 3-Way Contingency Table Based on ϕ-Divergence

In Chap. 3, we derived an approximation of the distribution of statistics used to test for independence in a 2-way contingency table based on an asymptotic expansion. On the basis of the approximation, a transformed statistic is constructed. Transformed statistics increase the speed of convergence to a chi-square limiting distribution. In this section, we first consider a 3-way $J \times K \times L$ contingency table. Let $X_{jk\ell}$, $j = 1, \dots, J, k = 1, \dots, K, \ell = 1, \dots, L$ be the frequency of cell (j, k, ℓ) and assume that the sum n of all frequencies is fixed. A generic $J \times K \times L$ contingency table is shown in Table 4.1.

Assume that the random vector $X = (X_{111}, \dots, X_{11L}, \dots, X_{1K1}, \dots, X_{1KL}, \dots, X_{J11}, \dots, X_{J1L}, \dots, X_{JK1}, \dots, X_{JKL})^\top$ is distributed according to a multinomial distribution $\mathcal{M}_{JKL}(n, p)$, where $p = (p_{111}, \dots, p_{11L}, \dots, p_{1K1}, \dots, p_{1KL}, \dots, p_{J11}, \dots, p_{J1L}, \dots, p_{JK1}, \dots, p_{JKL})^\top$, $0 < p_{jk\ell} < 1$, $j = 1, \dots, J, k = 1, \dots, K, \ell = 1, \dots, L$ and $\sum_{j=1}^{J} \sum_{k=1}^{K} \sum_{\ell=1}^{L} p_{jk\ell}=1$. For any $j = 1, \dots, J, k = 1, \dots, K, \ell = 1, \dots, L$, we define the marginal probabilities and the marginal frequencies of rows, columns and layers as follows.

This chapter is based on the following article: Japanese Journal of Statistics and Data Science, 4, Taneichi N., Sekiya Y. and Toyama J., Improvement of the test of independence among groups of factors in a multi-way contingency table, 181–213, Springer Nature (2021).

N. Taneichi and Y. Sekiya, *Improving Tests for Discrete Small Sample Data*, JSS Research Series in Statistics,
https://doi.org/10.1007/978-981-95-5301-3_4

Table 4.1 A generic $J \times K \times L$ contingency table

	Table of 1st layer			...	Table of Lth layer			Total
	X_{111}	$\cdots$	X_{1K1}	$\cdots$	X_{11L}	$\cdots$	X_{1KL}	$X_{1\cdot\cdot}$
	$\vdots$		$\vdots$	$\cdots$	$\vdots$		$\vdots$	$\vdots$
	X_{j11}	$\cdots$	X_{jK1}	$\cdots$	X_{j1L}	$\cdots$	X_{jKL}	$X_{j\cdot\cdot}$
	$\vdots$		$\vdots$	$\cdots$	$\vdots$		$\vdots$	$\vdots$
	X_{J11}	$\cdots$	X_{JK1}	$\cdots$	X_{J1L}	$\cdots$	X_{JKL}	$X_{J\cdot\cdot}$
Total	$X_{\cdot 11}$	$\cdots$	$X_{\cdot K1}$	$\cdots$	$X_{\cdot 1L}$	$\cdots$	$X_{\cdot KL}$	n

$$p_{j\cdot\cdot} = \sum_{k=1}^{K}\sum_{\ell=1}^{L} p_{jk\ell},\ p_{\cdot k\cdot} = \sum_{j=1}^{J}\sum_{\ell=1}^{L} p_{jk\ell},\ p_{\cdot\cdot\ell} = \sum_{j=1}^{J}\sum_{k=1}^{K} p_{jk\ell},$$

$$p_{jk\cdot} = \sum_{\ell=1}^{L} p_{jk\ell},\ p_{j\cdot\ell} = \sum_{k=1}^{K} p_{jk\ell},\ p_{\cdot k\ell} = \sum_{j=1}^{J} p_{jk\ell},$$

$$X_{j\cdot\cdot} = \sum_{k=1}^{K}\sum_{\ell=1}^{L} X_{jk\ell},\ X_{\cdot k\cdot} = \sum_{j=1}^{J}\sum_{\ell=1}^{L} X_{jk\ell},\ X_{\cdot\cdot\ell} = \sum_{j=1}^{J}\sum_{k=1}^{K} X_{jk\ell},$$

$$X_{jk\cdot} = \sum_{\ell=1}^{L} X_{jk\ell},\ X_{j\cdot\ell} = \sum_{k=1}^{K} X_{jk\ell},\ X_{\cdot k\ell} = \sum_{j=1}^{J} X_{jk\ell}.$$

We consider the following null hypotheses of 3-way contingency tables.

$$H_0^{(1)} : p_{jk\ell} = p_{j\cdot\cdot}\,p_{\cdot k\cdot}\,p_{\cdot\cdot\ell},\quad j = 1,\ldots,J,\ k = 1,\ldots,K,\ \ell = 1,\ldots,L. \tag{4.1}$$

$$H_0^{(2)} : p_{jk\ell} = p_{jk\cdot}\,p_{\cdot\cdot\ell},\quad j = 1,\ldots,J,\ k = 1,\ldots,K,\ \ell = 1,\ldots,L. \tag{4.2}$$

Equation (4.1) is the hypothesis of complete independence and (4.2) is the hypothesis of two factors independence from another.

Let $\hat{p}_{jk\ell}^{(r)}, r = 1, 2$ be the maximum likelihood estimators of $p_{jk\ell}$ under $H_0^{(r)}, r = 1, 2$, respectively. That is, $\hat{p}_{jk\ell}^{(1)} = X_{j\cdot\cdot}X_{\cdot k\cdot}X_{\cdot\cdot\ell}/n^3$ and $\hat{p}_{jk\ell}^{(2)} = X_{jk\cdot}X_{\cdot\cdot\ell}/n^2$. Let $\tilde{p}_{jk\ell}$ be the unrestricted maximum likelihood estimator of $p_{jk\ell}$, that is, $\tilde{p}_{jk\ell} = X_{jk\ell}/n$. Then, statistics based on ϕ-divergence for testing the hypotheses $H_0^{(r)}, r = 1, 2$ are given by

$$C_\phi^{(r)[3]} = 2n\sum_{j=1}^{J}\sum_{k=1}^{K}\sum_{\ell=1}^{L} \hat{p}_{jk\ell}^{(r)}\phi\left(\frac{\tilde{p}_{jk\ell}}{\hat{p}_{jk\ell}^{(r)}}\right),\quad r = 1, 2,$$

respectively, where ϕ is given in Chap. 2 (Zografos 1993; Pardo 2006).

When we choose a convex function ϕ^a given by (2.3) as ϕ, then for each $r = 1, 2$, $C_{\phi^a}^{(r)[3]}$ reduces to

$$R_{(r)}^{a[3]} \equiv C_{\phi^a}^{(r)[3]} = 2n \sum_{j=1}^{J} \sum_{k=1}^{K} \sum_{\ell=1}^{L} I^a(\tilde{p}_{jk\ell}, \hat{p}_{jk\ell}^{(r)}),\ r = 1, 2, \tag{4.3}$$

where $I^a(e, f)$ is given by (3.4).

The statistics $R_{(r)}^{a[3]}, r = 1, 2$ are based on power divergence (Cressie and Read (1984)), while $R_{(r)}^{0[3]}, r = 1, 2$ are the log likelihood ratio statistics, and $R_{(r)}^{1[3]}, r = 1, 2$ are Pearson's X^2 statistics. $R_{(r)}^{2/3[3]}, r = 1, 2$ are statistics that correspond to those recommended in Cressie and Read (1984) for goodness-of-fit testing. Under $H_0^{(r)}, r = 1, 2$, it is known that the statistics $C_{\phi}^{(r)[3]}, r = 1, 2$ have a chi-square limiting distribution. Degrees of freedom are $JKL - J - K - L + 2$ when $r = 1$ and $(JK - 1)(L - 1)$ when $r = 2$. The limiting distribution of test statistics for a contingency table can be found in Pardo (2006).

4.2 An Approximation for $C_{\phi}^{(1)[3]}$ Under $H_0^{(1)}$ Based on an Asymptotic Expansion

In this section, we first derive a local Edgeworth approximation for the probability of X under the null hypothesis $H_0^{(1)}$ given by (4.1).

For $j = 1, \ldots, J, k = 1, \ldots, K, \ell = 1, \ldots, L$, let $U_{jk\ell} = n^{-1/2}(X_{jk\ell} - nq_{jk\ell})$, where $q_{jk\ell} = p_{j\cdot\cdot}p_{\cdot k\cdot}p_{\cdot\cdot\ell}$. Then $U = (U_{111}, \ldots, U_{11L}, \ldots, U_{1K1}, \ldots, U_{1KL}, \ldots, U_{J11}, \ldots, U_{J1L}, \ldots, U_{JK1}, \ldots, U_{JK,L-1})^{\top}$ is a lattice random vector that takes values in the set

$$S^{[3]} = \{u = (u_{111}, \ldots, u_{11L}, \ldots, u_{1K1}, \ldots, u_{1KL}, \ldots, u_{J11}, \ldots, u_{J1L}, \ldots, u_{JK1}, \ldots, u_{JK,L-1})^{\top} : u = n^{-1/2}(x - np_0),\ x \in S_0^{[3]}\},$$

where

$$p_0 = (I_{N^{[3]}-1}\ O_{N^{[3]}-1})p_0^*, \tag{4.4}$$

with $p_0^* = (q_{111}, \ldots, q_{11L}, \ldots, q_{1K1}, \ldots, q_{1KL}, \ldots, q_{J11}, \ldots, q_{J1L}, \ldots, q_{JK1}, \ldots, q_{JKL})^{\top}$, $I_{N^{[3]}-1}$ being an $(N^{[3]} - 1) \times (N^{[3]} - 1)$ identity matrix, $O_{N^{[3]}-1}$ being an $(N^{[3]} - 1)$-dimensional column zero vector,

$$N^{[3]} = JKL, \tag{4.5}$$

and

$$S_0^{[3]} = \Bigg\{x = (x_{111}, \ldots, x_{11L}, \ldots, x_{1K1}, \ldots, x_{1KL}, \ldots, x_{J11}, \ldots, x_{J1L}, \\ \ldots, x_{JK1}, \ldots, x_{JK,L-1})^\top : x_{jk\ell} \text{ being non-negative integers and} \\ \sum_{j=1}^{J}\sum_{k=1}^{K}\sum_{\ell=1}^{L} x_{jk\ell} \le n + x_{JKL}\Bigg\}. \tag{4.6}$$

The following theorem gives a local Edgeworth expansion.

Theorem 4.1 *Let* $u = n^{-1/2}(x - np_0)$ *for each* $x \in S_0^{[3]}$. *Then*

$$\Pr\{U = u|H_0^{(1)}\} = n^{-(N^{[3]}-1)/2} g^{[3]}(u)\{1 + n^{-1/2}h_1^{[3]}(u) + n^{-1}h_2^{[3]}(u) \\ + n^{-3/2}h_3^{[3]}(u) + O(n^{-2})\},$$

where

$$g^{[3]}(u) = (2\pi)^{-(N^{[3]}-1)/2}|\Omega^{[3]}|^{-1/2}\exp\left\{-\frac{1}{2}u^\top(\Omega^{[3]})^{-1}u\right\},$$

$$\Omega^{[3]} = \mathrm{diag}(q_{111}, \ldots, q_{JK,L-1}) - p_0 p_0^\top,$$

$$h_1^{[3]}(u) = -\frac{1}{2}L_1^{1[3]} + \frac{1}{6}L_2^{3[3]},$$

$$h_2^{[3]}(u) = \frac{1}{2}\{h_1^{[3]}(u)\}^2 + \frac{1}{12}(1 - L_1^{0[3]}) + \frac{1}{4}L_2^{2[3]} - \frac{1}{12}L_3^{4[3]},$$

$$h_3^{[3]}(u) = -\frac{1}{3}\{h_1^{[3]}(u)\}^3 + h_1^{[3]}(u)h_2^{[3]}(u) + \frac{1}{12}L_2^{1[3]} - \frac{1}{6}L_3^{3[3]} + \frac{1}{20}L_4^{5[3]},$$

where p_0 *is given by (4.4),* $u_{JKL} = -(u_{111} + \cdots + u_{JK,L-1})$ *and*

$$L_b^{a[3]} = \sum_{j=1}^{J}\sum_{k=1}^{K}\sum_{\ell=1}^{L}\frac{u_{jk\ell}^a}{q_{jk\ell}^b}. \tag{4.7}$$

Theorem 4.1 is a special case of Theorem 4.3 given in Sect. 4.7 when $M = 3$. Proof of the Theorem 4.3 is shown in Taneichi et al. (2021).

For the same reason as that for test statistic $C_\phi^{(I)}$ for independence of a 2-way contingency table, we propose the use of J_1^* as an approximation for the distribution of $C_\phi^{(1)[3]}$ under $H_0^{(1)}$, that is, $\Pr\{C_\phi^{(1)[3]} \le b|H_0^{(1)}\} \approx J_1^*$, where J_1^* comes from a multivariate Edgeworth expansion. Hereafter, throughout this book, when we consider an approximation based on an asymptotic expansion, we use the approximation that comes from a multivariate Edgeworth expansion J_1^*. With regard to evaluation of J_1^*, we obtain the following theorem.

Theorem 4.2 *If we assume that* ϕ *is six times continuously differentiable, then the* J_1^* *term is approximated up to order* n^{-1} *as follows:*

$$J_1^* = \Pr\{\chi^2_{f^{[3]}} \le z\} + n^{-1}\sum_{\nu=0}^{3} d_\nu^{\phi[3]} \Pr\{\chi^2_{f^{[3]}+2\nu} \le z\} + O(n^{-2}), \tag{4.8}$$

where

$$\begin{aligned}
d_0^{\phi[3]} &= -\frac{1}{24}F_4^{[3]},\\
d_1^{\phi[3]} &= \frac{1}{24}[F_1^{[3]}\phi^{(4)}(1) + F_2^{[3]}\{\phi'''(1)+1\}^2 + (2F_1^{[3]} + F_3^{[3]})\phi'''(1)\\
&\quad +(F_3^{[3]} + F_4^{[3]})],\\
d_2^{\phi[3]} &= \frac{1}{24}[-F_1^{[3]}\phi^{(4)}(1) - 2F_2^{[3]}\{\phi'''(1)+1\}^2 - (2F_1^{[3]} + F_3^{[3]})\phi'''(1) - F_3^{[3]}],\\
d_3^{\phi[3]} &= \frac{1}{24}F_2^{[3]}\{\phi'''(1)+1\}^2,
\end{aligned}$$

$$F_1^{[3]} = -3(2G_2^{[3]} - G_3^{[3]} + G_4^{[3]}), \quad F_2^{[3]} = 3G_1^{[3]} + 6G_2^{[3]} - 4G_3^{[3]} + 6G_4^{[3]},$$

$$F_3^{[3]} = 2(G_3^{[3]} - 3G_4^{[3]}), \quad F_4^{[3]} = 2G_3^{[3]},$$

$$\begin{aligned}
G_1^{[3]} &= L_1^{0[3]} - (N^{[3]})^2\left(\frac{q_1^{[3]}}{J^2} + \frac{q_2^{[3]}}{K^2} + \frac{q_3^{[3]}}{L^2}\right) + 2(N^{[3]})^2,\\
G_2^{[3]} &= L_1^{0[3]} - N^{[3]}\left(\frac{q_1^{[3]}}{J} + \frac{q_2^{[3]}}{K} + \frac{q_3^{[3]}}{L}\right) + 2N^{[3]},\\
G_3^{[3]} &= L_1^{0[3]} - (q_1^{[3]} + q_2^{[3]} + q_3^{[3]}) + 2,\\
G_4^{[3]} &= 2(JK + JL + KL) - 4(J + K + L) + 6,
\end{aligned}$$

where

$$q_1^{[3]} = \sum_{j=1}^{J} p_{j\cdot\cdot}^{-1}, \quad q_2^{[3]} = \sum_{k=1}^{K} p_{\cdot k\cdot}^{-1}, \quad q_3^{[3]} = \sum_{\ell=1}^{L} p_{\cdot\cdot\ell}^{-1},$$

$f^{[3]} = N^{[3]} - J - K - L + 2$, $N^{[3]}$ *is given by (4.5),* $L_b^{a[3]}$ *is given by (4.7), and* χ_f^2 *denotes a chi-square random variable with degrees of freedom* f.

Theorem 4.2 is a special case of Theorem 4.4 given in Sect. 4.7 when $M = 3$. Proof of Theorem 4.4 is shown in Appendix A.4.

4.3 A Transformed Test Statistic for Complete Independence in a 3-Way Contingency Table

Applying the (2.10), (2.11), and (2.12) to evaluation (4.8) given by Theorem 4.2, we construct a transformation for improving small sample accuracy of a chi-square approximation of the distribution of $C_\phi^{(1)}$ under $H_0^{(1)}$. For the statistic $C_\phi^{(1)[3]}$, based on ϕ that satisfies

$$\phi'''(1) = -1 \text{ and } \phi^{(4)}(1) = 2, \tag{4.9}$$

we propose the transformed statistic $C_\phi^{B[3]}=T_B(C_\phi^{(1)[3]}; d_0^{\phi[3]}, f^{[3]}, n)$, since $d_1^{\phi[3]} = -d_0^{\phi[3]}$, and $d_2^{\phi[3]} = d_3^{\phi[3]} = 0$ hold in (4.8). For the other statistics, we propose $C_\phi^{I[3]}=T_I(C_\phi^{(1)[3]}; d_0^{\phi[3]}, d_2^{\phi[3]}, d_3^{\phi[3]}, f^{[3]}, n)$ and $C_\phi^{CF[3]}= T_{CF}(C_\phi^{(1)[3]}; d_0^{\phi[3]}, d_1^{\phi[3]}, d_2^{\phi[3]}, f^{[3]}, n)$. However, when ϕ satisfies condition (4.9), we note that $C_\phi^{B[3]} = C_\phi^{CF[3]}$.

Now, $d_0^{\phi[3]}, \dots, d_3^{\phi[3]}$ include $G_1^{[3]}, \dots, G_4^{[3]}$, which are functions of unknown parameters $p_{j\cdot\cdot}$, $p_{\cdot k\cdot}$ and $p_{\cdot\cdot\ell}$ with $j = 1, \dots, J, k = 1, \dots, K, \ell = 1, \dots, L$. Then, in practical application, we substitute the maximum likelihood estimates $\hat{p}_{j\cdot\cdot} = x_{j\cdot\cdot}/n$, $\hat{p}_{\cdot k\cdot} = x_{\cdot k\cdot}/n$, and $\hat{p}_{\cdot\cdot\ell} = x_{\cdot\cdot\ell}/n$ for $p_{j\cdot\cdot}$, $p_{\cdot k\cdot}$, and $p_{\cdot\cdot\ell}$, respectively, where $x_{j\cdot\cdot}$, $x_{\cdot k\cdot}$ and $x_{\cdot\cdot\ell}$ are observed values of $X_{j\cdot\cdot}$, $X_{\cdot k\cdot}$ and $X_{\cdot\cdot\ell}$, respectively.

In the case of the power divergence statistic $R_{(1)}^{a[3]} \equiv C_{\phi^a}^{(1)[3]}$, $R_B^{a[3]} \equiv C_{\phi^a}^{B[3]}$ is defined when $a = 0$ (the log likelihood ratio statistic), while $R_I^{a[3]} \equiv C_{\phi^a}^{I[3]}$ is defined when $a \neq 0$, and $R_{CF}^{a[3]} \equiv C_{\phi_a}^{CF[3]}$ is defined when both $a \neq 0$ and $a = 0$ and $R_{CF}^{0[3]} = R_B^{0[3]}$.

The approximation $\Pr\{C_\phi^{\xi[3]} < z\} \approx \Pr\{\chi^2_{f^{[3]}} < z\}$ ($\xi = B, I, CF$) is justified by (2.13). Therefore, for $\xi \in \{B, I, CF\}$,

$$\Pr\{C_\phi^{\xi[3]} \geq z\} \approx \Pr\{\chi^2_{f^{[3]}} \geq z\}$$

holds. This means that the p-values of the test statistics $C_\phi^{\xi[3]}$ with $\xi \in \{B, I, CF\}$ are approximated by $\Pr\{\chi^2_{f^{[3]}} \geq z\}$. The theory of the transformed statistics is extended to M-way contingency tables, where M ($M \geq 3$) is an arbitrary natural number, in Sect. 4.6.

4.4 Extension to a Test of Independence Among Groups of Factors from a Test of Complete Independence

In a 3-way contingency table, Kobe et al. (2015) derived approximations for the distributions of statistics for testing the hypothesis $H_0^{(2)}$ of one factor independence from the other two based on an asymptotic expansion. On the basis of the approximation, they constructed a transformed statistic that increases the speed of convergence to a chi-square distribution.

In a 3-way contingency table, we derive an approximation for the distribution of statistics for testing $H_0^{(1)}$ of complete independence based on an asymptotic expansion. On the basis of the approximation, we construct Bartlett adjustment, an improved transformed statistic and a Bartlett-type transformed statistic that increases the speed of convergence to a chi-square distribution.

Incidentally, by putting $i = (j-1)K + k,\ j = 1, \cdots, J,\ k = 1, \cdots, K$ in cell probability $p_{jk\ell}$ of a 3-way $J \times K \times L$ contingency table and considering $p_{i\ell},\ i = 1, \cdots, JK, \ell = 1, \cdots, L$, the model of one factor independence from the other two in a 3-way $J \times K \times L$ contingency table given by (4.2) is considered as a model of independence in a 2-way $JK \times L$ contingency table. Therefore, the results reported by Kobe et al. (2015) can be obtained by a simple modification of the results for independence of a 2-way contingency table defined in Chap. 3. Furthermore, for example, approximations for the distributions of statistics on the basis of an asymptotic expansion for testing the hypothesis of one factor independence from the other three in a 4-way $I \times J \times K \times L$ contingency table,

$$\begin{aligned} H_0^{(3)} &: p_{ijk\ell} = p_{i\cdots}p_{\cdot jk\ell}, \\ &\qquad i = 1, \ldots, I, j = 1, \ldots, J, k = 1, \ldots, K, \ell = 1, \ldots, L, \end{aligned}$$

or the hypothesis of independence of two factors from the other three in a 5-way $H \times I \times J \times K \times L$ contingency table,

$$\begin{aligned} H_0^{(4)} &: p_{hijk\ell} = p_{hi\cdots}p_{\cdot\cdot jk\ell}, \\ &\qquad h = 1, \ldots, H, i = 1, \ldots, I, j = 1, \ldots, J, k = 1, \ldots, K, \ell = 1, \ldots, L, \end{aligned}$$

can also be obtained by a simple modification of results for independence of a 2-way contingency table. Generally, approximation for the distribution of statistics on the basis of an asymptotic expansion for testing the hypothesis of independence of N $(M > N)$ factors from the other $M - N$ factors in an M-way contingency table can be obtained by a simple modification of results for independence of a 2-way contingency table.

Similarly, by a simple modification of the approximation for the distribution of statistics for testing the hypothesis of complete independence $H_0^{(1)}$, which was derived in Sect. 4.2, we can obtain the approximation for the distribution of statistics for testing, for example, the hypothesis of independence among two factors, two factors and one factor,

$$\begin{aligned} H_0^{(5)} &: p_{hijk\ell} = p_{hi\cdots}p_{\cdot\cdot jk\cdot}p_{\cdots\cdot\ell}, \\ &\qquad h = 1, \ldots, H, i = 1, \ldots, I, j = 1, \ldots, J, k = 1, \ldots, K, \ell = 1, \ldots, L, \end{aligned}$$

in a 5-way contingency table on the basis of an asymptotic expansion. Therefore, we also obtain transformed statistics based on the approximation, which was derived in Sect. 4.3. We can also obtain the approximation for the distribution of statistics for testing, for example, the hypothesis of independence among three factors, one factor and one factor,

$$\begin{aligned} H_0^{(6)} &: p_{hijk\ell} = p_{hij\cdot\cdot}p_{\cdots k\cdot}p_{\cdots\cdot\ell}, \\ &\qquad h = 1, \ldots, H, i = 1, \ldots, I, j = 1, \ldots, J, k = 1, \ldots, K, \ell = 1, \ldots, L, \end{aligned}$$

in a 5-way contingency table on the basis of an asymptotic expansion and transformed statistics based on the approximation.

4.5 Application to Real Data in Sect. 1.1

We test the hypothesis of independence of age, employment, and pair (Response, Mode) for the data in Table 1.1 in Sect. 1.1.

By putting $h = (i-1)J + j,\ i = 1, \ldots, I,\ j = 1, \ldots, J$ in cell probability of a 4-way $I \times J \times K \times L$ contingency table and considering $p_{hk\ell},\ h = 1, \ldots, IJ, k = 1, \ldots, K, \ell = 1, \ldots, L$, the model of one factor and one factor independence from the other two factors in a $I \times J \times K \times L$ contingency table given by Table 1.1 is considered as a model of complete independence in a 3-way $IJ \times K \times L$ contingency table. We also make $X_{hk\ell}$ corresponding to $p_{hk\ell}$ from a 4-way contingency table. By applying $p_{hk\ell}$ to the hypothesis of complete independence in a 3-way contingency table given by (4.1) and by applying $X_{hk\ell},\ h = 1, \ldots, IJ, k = 1, \ldots, K, \ell = 1, \ldots, L$ to statistic $R^{a[3]}_{(1)}$ given by (4.3), $R^{a[3]}_{(1)}$ is asymptotically distributed as a chi-square distribution with degrees of freedom $(IJ)KL - IJ - K - L + 2$. Also, we can obtain the transformed statistics $R^{a[3]}_{CF}$ by (2.12).

For the data in Table 1.1, the values of the statistics $R^{a[3]}_{(1)}$ and the transformed statistics $R^{a[3]}_{CF}$ for $a = 0, 0.2, 2/3, 1$ are listed in Table 4.2. Also, for the data in Table 1.1, the values of the asymptotic p-value (a.p-value) of $R^{a[3]}_{(1)}$ and $R^{a[3]}_{CF}$ and the values of the simulated p-value (s.p-value), which is more exact than the asymptotic p-value, for $a = 0, 0.2, 2/3, 1$ are listed in Table 4.3.

Table 4.2 Values of $R^{a[3]}_{(1)}$ and $R^{a[3]}_{CF}$ (a=0, 0.2, 2/3, 1) for the data in Table 1.1

a	value of $R^{a[3]}_{(1)}$	value of $R^{a[3]}_{CF}$
0.0	41.449	39.028
0.2	40.035	38.848
2/3	38.211	38.360
1.0	37.629	37.755

Table 4.3 Asymptotic and simulated p-values of $R^{a[3]}_{(1)}$ and $R^{a[3]}_{CF}$ (a=0, 0.2, 2/3, 1) for the data in Table 1.1

a	a.p-value of $R^{a[3]}_{(1)}$	s.p-value of $R^{a[3]}_{(1)}$	a.p-value of $R^{a[3]}_{CF}$	s.p-value of $R^{a[3]}_{CF}$
0.0	0.04882	0.08344	0.08047	0.08344
0.2	0.06566	0.08145	0.08340	0.08145
2/3	0.09447	0.08940	0.09178	0.08940
1.0	0.10563	0.10351	0.10313	0.10351

From Table 4.2, the value of the log likelihood ratio statistic $R_{(1)}^{0[3]} = G^2$ is 41.449. The value of the transformed log likelihood ratio statistic $R_B^{0[3]} = R_{CF}^{0[3]}$ is 39.028. Since $\chi^2_{28}(0.05) = 41.337$, the hypothesis that age, employment and pair (Response, Mode) are independent in the data in Table 1.1 cannot be rejected at the 5% level using the transformed statistics $R_B^{0[3]} = R_{CF}^{0[3]}$. From Table 4.3, the asymptotic p-value of $R_{(1)}^{0[3]} = G^2$ is 0.04882 and the simulated p-value, which is more exact than the asymptotic p-value, is 0.08344. The asymptotic p-value of $R_B^{0[3]} = R_{CF}^{0[3]}$ is 0.08047, and the simulated p-value of $R_B^{0[3]} = R_{CF}^{0[3]}$ is 0.08344. Therefore, for the statistic $R_{(1)}^{0[3]} = G^2$, the conclusion of the test based on an asymptotic distribution is opposite to that of the test based on a more exact distribution at the 5% level. In contrast, for the statistic $R_B^{0[3]} = R_{CF}^{0[3]}$, the conclusion of the test based on an asymptotic distribution agrees with that of the test based on a more exact distribution at the 5% level.

4.6 A Test Statistic for Complete Independence in an M-Way Contingency Table

In this section, the results derived in Sect. 4.2 are extended to an M-way contingency table. By this extension, we obtain transformed statistics for testing various kinds of independence among groups of factors in an M-way contingency table and we also obtain transformed statistics for testing complete independence of an M-way contingency table. We derive statistics for testing the complete independence in an M-way contingency table. First, we give notations and symbols as follows.

For an arbitrary positive integer M $(M \geq 3)$, we consider M-way $J_1 \times \cdots \times J_M$ contingency tables as a full multinomial model in which the sample size n is fixed. That is, let the number of the frequency of a cell $(j_1, \ldots, j_M)$, $j_m = 1, \ldots, J_m$; $m = 1, \ldots, M$, be $X_{j_1 \cdots j_M}$, where

$$\sum_{j_1=1}^{J_1} \cdots \sum_{j_M=1}^{J_M} X_{j_1 \cdots j_M} = n,$$

and let

$$X^* = (X_{1 \cdots 11}, \ldots, X_{1 \cdots 1 J_M}, \ldots, X_{J_1 \cdots J_M})^\top.$$

Then X^* is distributed according to $\mathcal{M}_{N^{[M]}}(n, p^*)$, where

$$N^{[M]} = \prod_{m=1}^{M} J_m, \tag{4.10}$$

$$p^* = (p_{1 \cdots 11}, \ldots, p_{1 \cdots 1 J_M}, \ldots, p_{J_1 \cdots J_M})^\top,$$

$$0 < p_{j_1 \cdots j_M} < 1, \quad j_m = 1, \dots, J_m, \ m = 1, \dots, M$$

and

$$\sum_{j_1=1}^{J_1} \cdots \sum_{j_M=1}^{J_M} p_{j_1 \cdots j_M} = 1.$$

Here, we define some symbols.

$$a_{\cdot(m, j_m)} = \sum_{j_1=1}^{J_1} \cdots \sum_{j_{m-1}=1}^{J_{m-1}} \sum_{j_{m+1}=1}^{J_{m+1}} \cdots \sum_{j_M=1}^{J_M} a_{j_1 \cdots j_m \cdots j_M}, \quad j_m = 1, \dots, J_m; \ m = 1, \dots, M,$$

where $a_{j_1 \cdots j_m \cdots j_M}$, $j_m = 1, \dots, J_m$; $m = 1, \dots, M$ is a sequence that has M indices. Similarly,

$$a_{\cdot(\ell, j_\ell; m, j_m)} = \sum_{j_1=1}^{J_1} \cdots \sum_{j_{\ell-1}=1}^{J_{\ell-1}} \sum_{j_{\ell+1}=1}^{J_{\ell+1}} \cdots \sum_{j_{m-1}=1}^{J_{m-1}} \sum_{j_{m+1}=1}^{J_{m+1}} \cdots \sum_{j_M=1}^{J_M} a_{j_1 \cdots j_\ell \cdots j_m \cdots j_M}, \quad j_\ell = 1, \dots, J_\ell, \ j_m = 1, \dots, J_m; \ 1 \le \ell < m \le M$$

and

$$a_{\cdot(k, j_k; \ell, j_\ell; m, j_m)} = \sum_{j_1=1}^{J_1} \cdots \sum_{j_{k-1}=1}^{J_{k-1}} \sum_{j_{k+1}=1}^{J_{k+1}} \cdots \sum_{j_{\ell-1}=1}^{J_{\ell-1}} \sum_{j_{\ell+1}=1}^{J_{\ell+1}} \cdots \sum_{j_{m-1}=1}^{J_{m-1}} \times \sum_{j_{m+1}=1}^{J_{m+1}} \cdots \sum_{j_M=1}^{J_M} a_{j_1 \cdots j_k \cdots j_\ell \cdots j_m \cdots j_M},$$

$$j_k = 1, \dots, J_k, \ j_\ell = 1, \dots, J_\ell, \ j_m = 1, \dots, J_m; \ 1 \le k < \ell < m \le M.$$

Symbols $p_{\cdot(m, j_m)}$, $X_{\cdot(m, j_m)}$, $U_{\cdot(m, j_m)}$, $u_{\cdot(m, j_m)}$, $u_{\cdot(\ell, j_\ell; m, j_m)}$, and $u_{\cdot(k, j_k; \ell, j_\ell; m, j_m)}$ used below are defined in the same way as $a_{\cdot(m, j_m)}$, $a_{\cdot(\ell, j_\ell; m, j_m)}$ and $a_{\cdot(k, j_k; \ell, j_\ell; m, j_m)}$. Symbols $p_{\cdot(1, j_1)}$ and $p_{\cdot(1, j_1; 2, j_2)}$ are extensions of symbols $p_{j\cdot\cdot}$ and $p_{jk\cdot}$ in a 3-way contingency table, respectively.

In this section, we give the test of the hypothesis for complete independence of M-way contingency tables as follows:

$$H_0^M : p_{j_1 \cdots j_M} = Q(j_1, \dots, j_M), \quad j_m = 1, \dots, J_m; \ m = 1, \dots, M, \tag{4.11}$$

where

$$Q(j_1, \dots, j_M) = \prod_{m=1}^{M} p_{\cdot(m, j_m)}, \quad j_m = 1, \dots, J_m; \ m = 1, \dots, M. \tag{4.12}$$

Hypothesis H_0^M means the hypothesis $H_0^{(1)}$ when $M = 3$. Statistic $C_\phi^{(1)[M]}$ based on ϕ-divergence for testing of a hypothesis H_0^M of complete independence given by (4.11) is defined by

$$C_\phi^{(1)[M]} = 2n \sum_{j_1=1}^{J_1} \cdots \sum_{j_M=1}^{J_M} \hat{Q}(j_1, \ldots, j_M)\phi\left(\frac{\hat{p}_{j_1\cdots j_M}}{\hat{Q}(j_1, \ldots, j_M)}\right), \tag{4.13}$$

where

$$\hat{Q}(j_1, \ldots, j_M) = \prod_{m=1}^{M} \hat{p}_{\cdot(m,j_m)}, \tag{4.14}$$

$$\hat{p}_{j_1\cdots j_M} = X_{j_1\cdots j_M}/n, \ \ j_m = 1, \ldots, J_m;\ m = 1, \ldots, M, \tag{4.15}$$

and

$$\hat{p}_{\cdot(m,j_m)} = X_{\cdot(m,j_m)}/n, \ \ j_m = 1, \ldots, J_m;\ \ m = 1, \ldots, M.$$

$C_{\phi^a}^{(1)[M]}$ become power divergence statistics

$$R_{(1)}^{a[M]} \equiv C_{\phi^a}^{(1)[M]} = 2n \sum_{j_1=1}^{J_1} \cdots \sum_{j_M=1}^{J_M} I^a\left(\hat{p}_{j_1\cdots j_M}, \hat{Q}(j_1, \ldots, j_M)\right)$$

when ϕ^a is given by (2.3). Under the hypothesis H_0^M of complete independence given by (4.11), all members of the class of statistics $C_\phi^{(1)[M]}$ have a limiting central chi-square distribution with degrees of freedom

$$f^{[M]} = N^{[M]} - \sum_{m=1}^{M} J_m + M - 1, \tag{4.16}$$

where $N^{[M]}$ is defined by (4.10).

4.7 An Approximation for $C_\phi^{(1)[M]}$ Under H_0^M Based on an Asymptotic Expansion

First, we derive a local Edgeworth expansion under the null hypothesis H_0^M. Since we consider a null distribution, we assume that X^* is distributed according to a multinomial distribution $\mathcal{M}_{N^{[M]}}(n, p_0^*)$, where p_0^* is

$$p_0^* = p_1 \otimes \cdots \otimes p_M, \tag{4.17}$$

where $p_m = (p_{\cdot(m,1)}, \ldots, p_{\cdot(m,J_m)})^\top$, $m = 1, \ldots, M$ and $\otimes$ denotes the Kronecker product. Let

$$U_{j_1 \cdots j_M} = n^{-1/2}\{X_{j_1 \cdots j_M} - nQ(j_1, \ldots, j_M)\}, \; j_m = 1, \ldots, J_m; \; m = 1, \ldots, M \tag{4.18}$$

and put $U^* = (U_{1\cdots 1}, \ldots, U_{J_1 \cdots J_M})^\top$. Let U be the $(N^{[M]} - 1)$-dimensional column vector that eliminates the last element of U^*. That is,

$$U = (U_{1\cdots 1}, \ldots, U_{J_1 \cdots J_{M-1} J_M - 1})^\top = (I_{N^{[M]}-1} \; O_{N^{[M]}-1})\, U^*,$$

where $I_{N^{[M]}-1}$ is an $(N^{[M]} - 1) \times (N^{[M]} - 1)$ identity matrix and $O_{N^{[M]}-1}$ is an $(N^{[M]} - 1)$-dimensional zero vector. Then, U is a lattice random vector that takes values in the set

$$S^{[M]} = \left\{ u : u = n^{-1/2}(x - np_0), \; x \in S_0^{[M]} \right\},$$

where

$$S_0^{[M]} = \Big\{ x = (x_{1\cdots 1}, \ldots, x_{J_1 \ldots J_{M-1} J_M - 1})^\top : x_{j_1 \ldots j_M} \text{ are non-negative integers and the sum of all elements of } x \text{ is less than or equal to } n \Big\}$$

and $p_0 = (I_{N^{[M]}-1} \; O_{N^{[M]}-1})\, p_0^*$. We obtain the following theorem for deriving a local Edgeworth expansion.

Theorem 4.3 *For each $x \in S_0^{[M]}$, let $u = n^{-1/2}(x - np_0)$, where $u = (u_{1\cdots 1}, \ldots, u_{J_1 \cdots J_{M-1} J_M - 1})^\top$. Then*

$$\Pr\{U = u \mid H_0^M\} = n^{-(N^{[M]}-1)/2} g^{[M]}(u) \Big\{ 1 + n^{-1/2} h_1^{[M]}(u) + n^{-1} h_2^{[M]}(u) + n^{-3/2} h_3^{[M]}(u) + O(n^{-2}) \Big\},$$

where

$$g^{[M]}(u) = (2\pi)^{-(N^{[M]}-1)/2} |\Omega^{[M]}|^{-1/2} \exp\left(-\frac{1}{2} u^\top (\Omega^{[M]})^{-1} u \right),$$

$$h_1^{[M]}(u) = -\frac{1}{2} L_1^{1[M]} + \frac{1}{6} L_2^{3[M]},$$

$$h_2^{[M]}(u) = \frac{1}{2} \left\{ h_1^{[M]}(u) \right\}^2 + \frac{1}{12} \left(1 - L_1^{0[M]} \right) + \frac{1}{4} L_2^{2[M]} - \frac{1}{12} L_3^{4[M]},$$

$$h_3^{[M]}(u) = -\frac{1}{3} \left\{ h_1^{[M]}(u) \right\}^3 + h_1^{[M]}(u) h_2^{[M]}(u) + \frac{1}{12} L_2^{1[M]} - \frac{1}{6} L_3^{3[M]} + \frac{1}{20} L_4^{5[M]},$$

$$\Omega^{[M]} = (I_{N^{[M]}-1}\ O_{N^{[M]}-1})\,\Omega^* \begin{pmatrix} I_{N^{[M]}-1} \\ O_{N^{[M]}-1}^\top \end{pmatrix},$$

$$\Omega^* = D_1 \otimes \cdots \otimes D_M - p_0^* \left(p_0^*\right)^\top,$$

$$D_m = \mathrm{diag}(p_{\cdot(m,1)}, \ldots, p_{\cdot(m,J_m)}) \quad (m = 1, \ldots, M),$$

$$L_b^{a[M]} = \sum_{j_1=1}^{J_1} \cdots \sum_{j_M=1}^{J_M} \frac{u_{j_1\cdots j_M}^a}{Q(j_1, \ldots, j_M)^b},$$

where $u_{J_1\cdots J_M} = -(u_{1\cdots 1} + \cdots + u_{J_1\cdots J_{M-1}J_M-1})$, $Q(j_1, \ldots, j_M)$ *and* p_0^* *are defined in (4.12) and (4.17), respectively.*

Proof of Theorem 4.3 is shown in Taneichi et al. (2021).

We consider the J_1^* term, that is, a term of multivariate Edgeworth expansion for the distribution of test statistic $C_\phi^{(1)[M]}$. By the transformation (4.18), statistic $C_\phi^{(1)[M]}$ can be rewritten as

$$C_\phi^{(1)[M]}(U) = 2n \sum_{j_1=1}^{J_1} \cdots \sum_{j_M=1}^{J_M} Q(j_1, \ldots, j_M)\phi\left(A_{j_1\cdots j_M}\right) \prod_{m=1}^{M} \left(1 + \frac{n^{-1/2}U_{\cdot(m,j_m)}}{p_{\cdot(m,j_m)}}\right),$$

where

$$A_{j_1\cdots j_M} = \left\{1 + \frac{n^{-1/2}U_{j_1\cdots j_M}}{Q(j_1, \ldots, j_M)}\right\} \prod_{m=1}^{M} \left(1 + \frac{n^{-1/2}U_{\cdot(m,j_m)}}{p_{\cdot(m,j_m)}}\right)^{-1}.$$

Let $B_\phi(z)$ be a set defined by

$$B_\phi(z) = \{u : C_\phi^{(1)[M]}(u) \le z | H_0^M\}.$$

Since J_1^* is a term of multivariate Edgeworth expansion assuming a continuous distribution, the J_1^* term can be represented as

$$J_1^* = \int \cdots \int_{B_\phi(z)} g^{[M]}(u)\Big\{1 + n^{-1/2}h_1^{[M]}(u) + n^{-1}h_2^{[M]}(u) + n^{-3/2}h_3^{[M]}(u)\Big\}du + O(n^{-2}), \tag{4.19}$$

where $g^{[M]}$, $h_1^{[M]}$, $h_2^{[M]}$ and $h_3^{[M]}$ are denoted in Theorem 4.3. With regard to the evaluation of J_1^*, we obtain the following theorem.

Theorem 4.4 *If we assume that* ϕ *is six times continuously differentiable, then the* J_1^* *term is approximated up to order* n^{-1} *as follows:*

$$J_1^* = \Pr\{\chi^2_{f^{[M]}} \le z\} + n^{-1}\sum_{\nu=0}^{3} d_\nu^{\phi[M]} \Pr\{\chi^2_{f^{[M]}+2\nu} \le z\} + O(n^{-2}), \tag{4.20}$$

where

$$\begin{aligned}
d_0^{\phi[M]} &= -\frac{1}{24}F_4^{[M]},\\
d_1^{\phi[M]} &= \frac{1}{24}[F_1^{[M]}\phi^{(4)}(1) + F_2^{[M]}\{\phi'''(1)+1\}^2 + (2F_1^{[M]} + F_3^{[M]})\phi'''(1)\\
&\quad +(F_3^{[M]} + F_4^{[M]})],\\
d_2^{\phi[M]} &= \frac{1}{24}[-F_1^{[M]}\phi^{(4)}(1) - 2F_2^{[M]}\{\phi'''(1)+1\}^2 - (2F_1^{[M]} + F_3^{[M]})\phi'''(1)\\
&\quad -F_3^{[M]}],\\
d_3^{\phi[M]} &= \frac{1}{24}F_2^{[M]}\{\phi'''(1)+1\}^2,
\end{aligned}$$

$$F_1^{[M]} = -3(2G_2^{[M]} - G_3^{[M]} + G_4^{[M]}), \quad F_2^{[M]} = 3G_1^{[M]} + 6G_2^{[M]} - 4G_3^{[M]} + 6G_4^{[M]},$$

$$F_3^{[M]} = 2(G_3^{[M]} - 3G_4^{[M]}), \quad F_4^{[M]} = 2G_3^{[M]},$$

$$\begin{aligned}
G_1^{[M]} &= L_1^{0[M]} - (N^{[M]})^2\sum_{m=1}^{M}\frac{q_m^{[M]}}{J_m^2} + (M-1)(N^{[M]})^2,\\
G_2^{[M]} &= L_1^{0[M]} - N^{[M]}\sum_{m=1}^{M}\frac{q_m^{[M]}}{J_m} + (M-1)N^{[M]},\\
G_3^{[M]} &= L_1^{0[M]} - \sum_{m=1}^{M} q_m^{[M]} + (M-1),\\
G_4^{[M]} &= 2\sum_{m=1}^{M-1}\sum_{\ell=m+1}^{M} J_m J_\ell - 2(M-1)\sum_{m=1}^{M} J_m + M(M-1),
\end{aligned}$$

where

$$q_m^{[M]} = \sum_{j_m=1}^{J_m} p_{\cdot(m,j_m)}^{-1}, \quad m = 1, \ldots, M,$$

and $N^{[M]}$ and $f^{[M]}$ are defined in (4.10) and (4.16), respectively.

Proof of Theorem 4.4 is shown in Appendix A.4.

4.8 A Transformed Test Statistic for Complete Independence in an M-Way Contingency Table

By Theorem 4.4, we can extend the transformed statistic $C_{\phi}^{B[3]}$ to $C_{\phi}^{B[M]} = T_B(C_{\phi}^{(1)[M]}; d_0^{\phi[M]}, f^{[M]}, n)$, $M = 4, 5, \cdots$. Also, we can extend the transformed statistic $C_{\phi}^{I[3]}$ to $C_{\phi}^{I[M]} = T_I(C_{\phi}^{(1)[M]}; d_0^{\phi[M]}, d_2^{\phi[M]}, d_3^{\phi[M]}, f^{[M]}, n)$, $M = 4, 5, \cdots$ and the transformed statistic $C_{\phi}^{CF[3]}$ to $C_{\phi}^{CF[M]} = T_{CF}(C_{\phi}^{(1)[M]}; d_0^{\phi[M]}, d_1^{\phi[M]}, d_2^{\phi[M]}, f^{[M]}, n)$, $M = 4, 5, \cdots$.

Therefore, in the case of the power divergence statistic $R_{(1)}^{a[M]} = C_{\phi^a}^{(1)[M]}$, $R_B^{a[M]} = C_{\phi^a}^{B[M]}$ is obtained when $a = 0$ (the log likelihood ratio statistic), while $R_I^{a[M]} = C_{\phi^a}^{I[M]}$ is obtained when $a \neq 0$, and $R_{CF}^{a[M]} = C_{\phi^a}^{CF[M]}$ is obtained when both $a \neq 0$ and $a = 0$ and $R_{CF}^{0[M]} = R_B^{0[M]}$.

The approximation $\Pr\{C_{\phi}^{\xi[M]} < z\} \approx \Pr\{\chi^2_{f^{[M]}} < z\}$ $(\xi = B, I, CF)$, $M = 4, 5, \cdots$ is also justified by (2.13). Therefore, for $\xi \in \{B, I, CF\}$,

$$\Pr\{C_{\phi}^{\xi[M]} \geq z\} \approx \Pr\{\chi^2_{f^{[M]}} \geq z\}, \;\; M = 4, 5, \cdots$$

also holds. This means that the p-values of the test statistics $C_{\phi}^{\xi[M]}$ with $\xi \in \{B, I, CF\}$, $M = 4, 5, \cdots$ are also approximated by $\Pr\{\chi^2_{f^{[M]}} \geq z\}$.

In Taneichi et al. (2021), the small sample performance of approximation by a chi-square distribution of test statistics of complete independence of 3-way $2 \times 2 \times 2$, $2 \times 3 \times 4$, and $3 \times 3 \times 3$ contingency tables and 4-way $2 \times 2 \times 2 \times 2$, $2 \times 2 \times 3 \times 3$, and $2 \times 2 \times 3 \times 4$ contingency tables were investigated. For power divergence statistics, performance of $R_I^{a[c]}$ $(a = 0, 0.2, 2/3, 1)$, $R_{CF}^{a[c]}$ $(a = 0.2, 2/3, 1)$, and $R_B^{0[c]}$, $(c = 3, 4)$ were numerically compared with those of the original statistics $R^{a[c]}$ $(a = 0, 0.2, 2/3, 1)$, $(c = 3, 4)$ for significance levels 0.01 and 0.05. The results of comparisons were as follows. Transformed statistics $R_I^{a[c]}$ $(a = 0, 0.2)$, $R_{CF}^{a[c]}$ $(a = 0, 0.2)$ and $R_B^{0[c]}$, $(c = 3, 4)$ performed much better than the original statistics $R^{a[c]}$ $(a = 0, 0.2)$, $(c = 3, 4)$. However, performance of $R_I^{a[c]}$ $(a = 2/3, 1)$ and $R_{CF}^{a[c]}$ $(a = 2/3, 1)$, $(c = 3, 4)$ were not so better than that of the original statistics $R^{a[c]}$ $(a = 2/3, 1)$, $(c = 3, 4)$ as compared to the case of $a = 0$ and $a = 0.2$. Furthermore, in Taneichi et al. (2021), the power against some alternative hypotheses of transformed statistics $R_I^{a[c]}$ $(a = 0, 0.2, 2/3, 1)$, $R_{CF}^{a[c]}$ $(a = 0.2, 2/3, 1)$, and $R_B^{0[c]}$ $(c = 3, 4)$ were numerically compared with that of the original statistics $R^{a[c]}$ $(a = 0, 0.2, 2/3, 1)$ $(c = 3, 4)$. The results of comparisons were as follows. The power of the transformed statistics $R_I^{a[c]}$ $(a = 0, 0.2, 2/3, 1)$, $R_{CF}^{a[c]}$ $(a = 0.2, 2/3, 1)$, and $R_B^{0[c]}$, $(c = 3, 4)$ were almost the same as that of the original statistics $R^{a[c]}$ $(a = 0, 0.2, 2/3, 1)$, $(c = 3, 4)$.

References

Cressie N, Read TRC (1984) Multinomial goodness-of-fit tests. J Roy Statist Soc Ser B 46:440–464

Kobe T, Taneichi N, Sekiya Y (2015) Improved transformed statistics for the test of one factor independence from the other two in an $r \times s \times t$ contingency table. J Jpn Stat Soc 45(1):77–98

Pardo L (2006) Statistical inference based on divergence measures. Chapman & Hall/CRC

Taneichi N, Sekiya Y, Toyama J (2021) Improvement of the test of independence among groups of factors in a multi-way contingency table. Jpn J Stat Data Sci 4:181–213

Zografos K (1993) Asymptotic properties of Φ-divergence statistic and its application in contingency tables. Internat J Math Statist Sci 2:5–21

Chapter 5
A Transformed Statistic for the Test of Conditional Independence in a $J \times K \times L$ Table

Keywords Asymptotic expansion · Conditional independence · Local Edgeworth expansion · 3-way contingency table · transformed statistics

5.1 Statistic for the Test of Conditional Independence in a 3-Way Contingency Table Based on ϕ-Divergence

In this section, we consider the test statistic for conditional independence in a 3-way contingency table given by Table 4.1. Let $X_{jk\ell}$ be distributed according to $\mathcal{M}_{JKL}(n, p)$ as shown in Sect. 4.1 and let marginal probabilities $p_{j\cdot\cdot}$, $p_{\cdot k\cdot}$, $p_{\cdot\cdot\ell}$, $p_{jk\cdot}$, $p_{j\cdot\ell}$, and $p_{\cdot k\ell}$ also be given as shown in Sect. 4.1.

We derive the hypothesis of conditional independence in the $J \times K \times L$ contingency table. By the definition of conditional probability, when $\Pr\{\text{layer} = \ell\} > 0$, the probability of row j and column k given that the layer ℓ is

$$
\begin{aligned}
&\Pr\{\text{row} = j,\ \text{column} = k \mid \text{layer} = \ell\} \\
&= \frac{\Pr\{\text{row} = j,\ \text{column} = k,\ \text{layer} = \ell\}}{\Pr\{\text{layer} = \ell\}} \\
&= \frac{p_{jk\ell}}{p_{\cdot\cdot\ell}}.
\end{aligned}
\tag{5.1}
$$

When $\Pr\{\text{layer} = \ell\} > 0,\ \ell = 1, \ldots, L$, conditional independence of rows and columns for each layer means that for all j, k and ℓ,

This chapter is based on the following article: Journal of Multivariate Analysis, 171, Taneichi N., Sekiya Y. and Toyama J., Transformed statistics for tests of conditional independence in $J \times K \times L$ contingency tables, 193–208, Copyright Elsevier (2019).

N. Taneichi and Y. Sekiya, *Improving Tests for Discrete Small Sample Data*, JSS Research Series in Statistics,
https://doi.org/10.1007/978-981-95-5301-3_5

$$\begin{aligned}&\Pr\{\text{row} = j,\ \text{column} = k \mid \text{layer} = \ell\}\\&= \Pr\{\text{row} = j \mid \text{layer} = \ell\}\Pr\{\text{column} = k \mid \text{layer} = \ell\}\\&= \left(\frac{p_{j\cdot\ell}}{p_{\cdot\cdot\ell}}\right)\left(\frac{p_{\cdot k\ell}}{p_{\cdot\cdot\ell}}\right).\end{aligned} \tag{5.2}$$

From (5.1) and (5.2), when $p_{\cdot\cdot\ell} > 0,\ \ell = 1, \dots, L$, then the null hypothesis that rows and columns are independent given the layers is

$$H_{0C}^{(1)} : p_{jk\ell} = \frac{p_{j\cdot\ell}\,p_{\cdot k\ell}}{p_{\cdot\cdot\ell}},\quad j = 1, \dots, J,\ k = 1, \dots, K,\ \ell = 1, \dots, L. \tag{5.3}$$

Similarly, the null hypothesis that rows and layers are independent given columns is

$$H_{0C}^{(2)} : p_{jk\ell} = \frac{p_{jk\cdot}\,p_{\cdot k\ell}}{p_{\cdot k\cdot}},\quad j = 1, \dots, J,\ k = 1, \dots, K,\ \ell = 1, \dots, L, \tag{5.4}$$

and the null hypothesis that columns and layers are independent given rows is

$$H_{0C}^{(3)} : p_{jk\ell} = \frac{p_{jk\cdot}\,p_{j\cdot\ell}}{p_{j\cdot\cdot}},\quad j = 1, \dots, J,\ k = 1, \dots, K,\ \ell = 1, \dots, L. \tag{5.5}$$

The unrestricted maximum likelihood estimator of $p_{jk\ell}$ is $\tilde{p}_{jk\ell} = X_{jk\ell}/n$. Maximum likelihood estimators of $p_{jk\ell}$ under $H_{0C}^{(m)},\ m = 1, 2, 3$ are

$$\hat{p}_{jk\ell}^{(1)} = \frac{X_{j\cdot\ell}X_{\cdot k\ell}}{nX_{\cdot\cdot\ell}},\quad \hat{p}_{jk\ell}^{(2)} = \frac{X_{jk\cdot}X_{\cdot k\ell}}{nX_{\cdot k\cdot}}\quad \text{and}\quad \hat{p}_{jk\ell}^{(3)} = \frac{X_{jk\cdot}X_{j\cdot\ell}}{nX_{j\cdot\cdot}},$$

respectively, where $X_{j\cdot\cdot},\ X_{\cdot k\cdot},\ X_{\cdot\cdot\ell},\ X_{jk\cdot},\ X_{j\cdot\ell}$, and $X_{\cdot k\ell}$ are given in Sect. 4.1.

The ϕ-divergence statistics for testing $H_{0C}^{(m)},\ m = 1, 2, 3$ are

$$C_{\phi C}^{(m)} = 2n\sum_{j=1}^{J}\sum_{k=1}^{K}\sum_{\ell=1}^{L}\hat{p}_{jk\ell}^{(m)}\phi\left(\frac{\tilde{p}_{jk\ell}}{\hat{p}_{jk\ell}^{(m)}}\right),\quad m = 1, 2, 3,$$

where ϕ is given in Chap. 2 (Zografos 1993; Pardo 2006). When we choose a convex function ϕ^a given by (2.3) as ϕ, then for each $m = 1, 2, 3$, $C_{\phi^a C}^{(m)}$ reduces to

$$R_{(m)}^{a} \equiv C_{\phi^a C}^{(m)} = 2n\sum_{j=1}^{J}\sum_{k=1}^{K}\sum_{\ell=1}^{L} I^a(\tilde{p}_{jk\ell}, \hat{p}_{jk\ell}^{(m)}),\quad m = 1, 2, 3, \tag{5.6}$$

where $I^a(e, f)$ is given by (3.4). $R_{(m)}^{a},\ m = 1, 2, 3$ are statistics based on power divergence proposed by Cressie and Read (1984). $R_{(m)}^{0},\ m = 1, 2, 3$ are the log likelihood ratio statistics, and $R_{(m)}^{1},\ m = 1, 2, 3$ are Pearson's X^2 statistics. Statistics $R_{(m)}^{2/3},\ m = 1, 2, 3$ correspond to the statistics recommended by Cressie and Read

(1984) for the goodness-of-fit test. Under the null hypotheses $H_{0C}^{(m)}$, $m = 1, 2, 3$ given by (5.3), (5.4) and (5.5), it is known that all members of the class of statistics $C_{\phi C}^{(m)}$, $m = 1, 2, 3$ have a chi-square limiting distribution with degrees of freedom $(J-1)(K-1)L$, $(J-1)(L-1)K$ and $(K-1)(L-1)J$, respectively. The limiting distribution of test statistics for contingency tables can be found in Pardo (2006).

5.2 Local Edgeworth Expansion

We consider the null hypothesis of conditional independence. In this case, the structure of $H_{0C}^{(m)}$ and statistics $R_{(m)}^a$ and $C_{\phi C}^{(m)}$ for testing the hypothesis $H_{0C}^{(m)}$ are essentially the same for each $m = 1, 2, 3$. Therefore, it is sufficient to consider only $H_{0C}^{(1)}$, $R_{(1)}^a$ and $C_{\phi C}^{(1)}$. Hereafter, we write H_0, R^a and C_ϕ as $H_{0C}^{(1)}$, $R_{(1)}^a$ and $C_{\phi C}^{(1)}$, respectively. In this section, we derive a local Edgeworth approximation for the probability of X under the null hypothesis H_0. Let X be distributed according to $\mathcal{M}_{JKL}(n, p_0)$, where $p_0 = (q_{111}, \ldots, q_{11L}, \ldots, q_{1K1}, \ldots, q_{1KL}, \ldots, q_{J11}, \ldots, q_{J1L}, \ldots, q_{JK1}, \ldots, q_{JKL})^\top$ and

$$q_{jk\ell} = \frac{p_{j\cdot\ell}\, p_{\cdot k\ell}}{p_{\cdot\cdot\ell}}, \quad j = 1, \ldots, J, \; k = 1, \ldots, K, \; \ell = 1, \ldots, L.$$

Let

$$U_{jk\ell} = n^{-1/2}(X_{jk\ell} - nq_{jk\ell}), \quad j = 1, \ldots, J, \; k = 1, \ldots, K, \; \ell = 1, \ldots, L. \tag{5.7}$$

Then $U = (U_{111}, \ldots, U_{11L}, \ldots, U_{1K1}, \ldots, U_{1KL}, \ldots, U_{J11}, \ldots, U_{J1L}, \ldots, U_{JK1}, \ldots, U_{JK,L-1})^\top$ is a lattice random vector that takes values in the set

$$S = \Big\{u = (u_{111}, \ldots, u_{11L}, \ldots, u_{1K1}, \ldots, u_{1KL}, \ldots, u_{J11}, \ldots, u_{J1L}, \ldots, u_{JK1}, \ldots, u_{JK,L-1})^\top : u = n^{-1/2}(\tilde{x} - n\tilde{p}_0), \; \tilde{x} \in S_0^{[3]}\Big\},$$

where

$$\tilde{p}_0 = (I_{N-1} \; O_{N-1})p_0, \tag{5.8}$$

where I_{N-1} is an $(N-1) \times (N-1)$ identity matrix, O_{N-1} is an $(N-1)$-dimensional zero vector,

$$N = JKL, \tag{5.9}$$

and $S_0^{[3]}$ is given by (4.6). We obtain the following theorem for deriving a local Edgeworth expansion.

Theorem 5.1 *Let $u = n^{-1/2}(\tilde{x} - n\tilde{p}_0)$ for each $\tilde{x} \in S_0^{[3]}$. Then the probability that $U = u$ under H_0 is evaluated as*

$$\Pr\{U = u|H_0\} = n^{-(N-1)/2} f(u) \times \left\{1 + n^{-1/2}h_1(u) + n^{-1}h_2(u) + n^{-3/2}h_3(u) + O(n^{-2})\right\},$$

where

$$f(u) = (2\pi)^{-(N-1)/2}|\Omega|^{-1/2} \exp\left(-\frac{1}{2}u^{\top}\Omega^{-1}u\right), \tag{5.10}$$

$$\Omega = \text{diag}\left(q_{111}, \ldots, q_{JK,L-1}\right) - \tilde{p}_0\tilde{p}_0^{\top}, \tag{5.11}$$

$$h_1(u) = -\frac{1}{2}M_1^1 + \frac{1}{6}M_2^3, \tag{5.12}$$

$$h_2(u) = \frac{1}{2}\{h_1(u)\}^2 + \frac{1}{12}(1 - M_1^0) + \frac{1}{4}M_2^2 - \frac{1}{12}M_3^4, \tag{5.13}$$

$$h_3(u) = -\frac{1}{3}\{h_1(u)\}^3 + h_1(u)h_2(u) + \frac{1}{12}M_2^1 - \frac{1}{6}M_3^3 + \frac{1}{20}M_4^5, \tag{5.14}$$

where

$$M_b^a = \sum_{j=1}^{J}\sum_{k=1}^{K}\sum_{\ell=1}^{L}\frac{u_{jk\ell}^a}{q_{jk\ell}^b},$$

$$u_{JKL} = -(u_{111} + \cdots + u_{JK,L-1}),$$

and $\tilde{p}_0$ is given by (5.8).

The proof of Theorem 5.1 is shown in Taneichi et al. (2019).

5.3 An Approximation for C_ϕ Under H_0 Based on an Asymptotic Expansion

We consider the following approximation for the distribution of C_ϕ under H_0.

$$\Pr\{C_\phi \leq x|H_0\} \approx J_1^*,$$

where J_1^* comes from a multivariate Edgeworth expansion.

With regard to evaluation of the J_1^* term, we obtain the following theorem.

Theorem 5.2 *If we assume that ϕ is six times continuously differentiable, then the J_1^* term is evaluated as*

$$J_1^* = \Pr\{\chi_M^2 \leq x\} + n^{-1}\sum_{j=0}^{3} v_j^{\phi} \Pr\{\chi_{M+2j}^2 \leq x\} + O(n^{-2}), \tag{5.15}$$

where

$$
\begin{aligned}
v_0^\phi &= \gamma_0,\\
v_1^\phi &= (\gamma_1+\gamma_2)+2\phi'''(1)\gamma_2+\phi^{(4)}(1)\gamma_3+\{\phi'''(1)\}^2\gamma_4,\\
v_2^\phi &= 2(-\gamma_2+\gamma_3)-2\phi'''(1)(\gamma_2+\gamma_4)-\phi^{(4)}(1)\gamma_3-2\{\phi'''(1)\}^2\gamma_4,\\
v_3^\phi &= \gamma_4\{1+\phi'''(1)\}^2,
\end{aligned}
$$

$$
\begin{aligned}
\gamma_0 &= -\frac{1}{12}\Gamma_1-\frac{1}{12}\Gamma_2+\frac{1}{12}(\Gamma_3+\Gamma_4),\\
\gamma_1 &= \frac{1}{4}JK\Gamma_1+\frac{1}{4}\Gamma_2-\frac{1}{4}(K\Gamma_3+J\Gamma_4),\\
\gamma_2 &= \frac{1}{8}J^2K^2\Gamma_1+\frac{1}{8}\Gamma_2-\frac{1}{8}(K^2\Gamma_3+J^2\Gamma_4)\\
\gamma_3 &= \left(-\frac{1}{2}JK-\frac{1}{8}+\frac{1}{4}J+\frac{1}{4}K\right)\Gamma_1-\frac{1}{8}\Gamma_2+\frac{1}{4}(K\Gamma_3+J\Gamma_4)-\frac{1}{8}(\Gamma_3+\Gamma_4)\\
\gamma_4 &= \left(\frac{1}{8}J^2K^2+\frac{3}{4}JK+\frac{1}{3}-\frac{1}{2}J-\frac{1}{2}K\right)\Gamma_1+\frac{5}{24}\Gamma_2-\frac{1}{8}(K^2\Gamma_3+J^2\Gamma_4)\\
&\quad -\frac{1}{4}(K\Gamma_3+J\Gamma_4)+\frac{1}{6}(\Gamma_3+\Gamma_4),
\end{aligned}
$$

$$
\Gamma_1=\sum_{\ell=1}^{L}\frac{1}{p_{\cdot\cdot\ell}},\quad \Gamma_2=\sum_{j=1}^{J}\sum_{k=1}^{K}\sum_{\ell=1}^{L}\frac{1}{q_{jk\ell}},
$$

$$
\Gamma_3=\sum_{j=1}^{J}\sum_{\ell=1}^{L}\frac{1}{p_{j\cdot\ell}},\quad \Gamma_4=\sum_{k=1}^{K}\sum_{\ell=1}^{L}\frac{1}{p_{\cdot k\ell}},
$$

$M=(J-1)(K-1)L$, $q_{jk\ell}$ *is defined in Sect. 5.2,* and χ^2_ν *denotes a chi-square random variable with degrees of freedom* ν.

It is easily verified that $\sum_{j=0}^{3}v_j^\phi=0$. The proof of Theorem 5.2 is shown in Appendix **??**. By applying ϕ^a as ϕ in Theorem 5.2, we obtain the following corollary for the statistics based on power divergence.

Corollary 5.1 *When the statistic is* R^a *given by (5.6), the* J_1^* *term is evaluated as*

$$
J_1^*=\Pr\{\chi^2_M\le x\}+n^{-1}\sum_{j=0}^{3}v_j^{(a)}\Pr\{\chi^2_{M+2j}\le x\}+O(n^{-2}),
$$

where $v_j^{(a)},\ j=0,1,2,3$ *are defined as* $v_j^\phi,\ j=0,1,2,3$ *in the case of* $\phi'''(1)=a-1$ *and* $\phi^{(4)}(1)=(a-1)(a-2)$, *respectively.*

5.4 A Transformed Statistic for the Test of Conditional Independence Based on the Approximation

Applying (2.10), (2.11), and (2.12) to the evaluation (5.15) given by Theorem 5.2, we construct a transformation for improving small sample accuracy of chi-square approximations of the distribution of C_ϕ under H_0.

For the statistic C_ϕ based on ϕ that satisfies (4.9), we propose the transformed statistic $C_\phi^B{=}T_B(C_\phi;\, v_0^\phi, M, n)$, since $v_1^\phi = -v_0^\phi$, and $v_2^\phi = v_3^\phi = 0$ hold in (5.15). For the other statistics, we propose $C_\phi^I = T_I(C_\phi;\, v_0^\phi, v_2^\phi, v_3^\phi, M, n)$ and $C_\phi^{CF} = T_{CF}(C_\phi;\, v_0^\phi, v_1^\phi, v_2^\phi, M, n)$. However, when ϕ satisfies (4.9), we note that $C_\phi^B = C_\phi^{CF}$.

Now, v_j^ϕ, $j = 0, 1, 2, 3$ include Γ_i, $i = 1, 2, 3, 4$, which are functions of unknown parameters $p_{\cdot\cdot\ell}$, $\ell = 1, \dots, L$, $p_{j\cdot\ell}$, $j = 1, \dots, J, \ell = 1, \dots, L$ and $p_{\cdot k\ell}$, $k = 1, \dots, K, \ell = 1, \dots, L$. Then, in practical application, we substitute the maximum likelihood estimates $\check{p}_{\cdot\cdot\ell} = x_{\cdot\cdot\ell}/n$, $\check{p}_{j\cdot\ell} = x_{j\cdot\ell}/n$, and $\check{p}_{\cdot k\ell} = x_{\cdot k\ell}/n$ for $p_{\cdot\cdot\ell}$, $p_{j\cdot\ell}$, and $p_{\cdot k\ell}$, respectively, where $x_{\cdot\cdot\ell}$, $x_{j\cdot\ell}$ and $x_{\cdot k\ell}$ are observed values of $X_{\cdot\cdot\ell}$, $X_{j\cdot\ell}$ and $X_{\cdot k\ell}$, respectively.

In the case of the power divergence statistic $R^a \equiv C_{\phi^a}$, $R_B^a \equiv C_{\phi^a}^B$ is defined when $a = 0$ (the log likelihood ratio statistic). $R_I^a \equiv C_{\phi^a}^I$ is defined when $a \neq 0$, and $R_{CF}^a \equiv C_{\phi^a}^{CF}$ is defined when both $a \neq 0$ and $a = 0$ and $R_{CF}^0 = R_B^0$.

Approximation $\Pr\{C_\phi^\xi < x\} \approx \Pr\{\chi_M^2 < x\}$ $(\xi = B, I, CF)$ is a well-grounded approximation, since Eq. (2.13) holds. Therefore,

$$\Pr\{C_\phi^\xi \geq x\} \approx \Pr\{\chi_M^2 \geq x\} \quad (\xi = B, I, CF)$$

holds. This means that the p-value of the test statistics C_ϕ^ξ $(\xi = B, I, CF)$ is approximated by $\Pr\{\chi_M^2 \geq x\}$.

Taneichi et al. (2019) investigated the performance of an approximation to a chi-square distribution of a test statistic of conditional independence of 3-way $2 \times 2 \times 2$, $2 \times 2 \times 4$, and $3 \times 3 \times 3$ contingency tables. For power divergence statistics, performance of transformed statistics R_I^a $(a = 0, 0.2, 2/3, 1)$, R_B^0, and R_{CF}^a $(a = 0.2, 2/3, 1)$ were numerically compared with that of the original statistics R^a $(a = 0, 0.2, 2/3, 1)$ for significance levels 0.01 and 0.05. The results of comparisons were as follows. The transformed statistics R_I^a $(a = 0, 0.2)$ and R_{CF}^a $(a = 0, 0.2)$ performed much better than the original statistics R^a $(a = 0, 0.2)$. The transformed statistics $R_{CF}^{2/3}$ and $R_I^{2/3}$ performed better than $R^{2/3}$ in most cases, and $R_{CF}^{2/3}$ performed better than $R_I^{2/3}$. The transformed statistic R_{CF}^1 performed better than Pearson's X^2 statistic R^1 in most cases. However, R_I^1 sometimes did not perform better than R^1. Furthermore, in Taneichi et al. (2019), the power against some alternative hypotheses of transformed statistics R_I^a $(a = 0, 0.2, 2/3, 1)$ and R_{CF}^a $(a = 0, 0.2, 2/3, 1)$ were numerically compared with those of the original statistics R^a $(a = 0, 0.2, 2/3, 1)$. The results of the comparisons were as follows. The power of the transformed statistics R_I^a $(a = 0, 0.2, 2/3, 1)$ and R_{CF}^a $(a = 0, 0.2, 2/3, 1)$ were almost the same as that of the original statistics R^a $(a = 0, 0.2, 2/3, 1)$.

References

Cressie N, Read TRC (1984) Multinomial goodness-of-fit tests. J Roy Statist Soc Ser B 46:440–464

Pardo L (2006) Statistical inference based on divergence measures. Chapman & Hall/CRC

Taneichi N, Sekiya Y, Toyama J (2019) Transformed statistics for tests of conditional independence in $J \times K \times L$ contingency tables. J Multivariate Anal 171:193–208

Zografos K (1993) Asymptotic properties of Φ-divergence statistic and its application in contingency tables. Internat J Math Statist Sci 2:5–21

Chapter 6
A Transformed Goodness-of-Fit Statistic for a Generalized Linear Model of Binary Data

Keywords Asymptotic expansion · Binary data · Generalized linear model · Local edgeworth expansion · Transformed statistics

6.1 Goodness-of-Fit Statistics for a Generalized Linear Model of Binary Data

We consider generalized linear models (Nelder and Wedderburn 1972) in which the response variables are measured on a binary scale. Let N independent random variables $Y_\alpha,\ \alpha = 1, \ldots, N$ corresponding to the number of successes in N different subgroups be distributed according to binomial distributions $B(n_\alpha, \pi_\alpha)$, $\alpha = 1, \ldots, N$. If we use a monotone and differentiable function g as a link function, we obtain a generalized linear model for binary data as follows.

$$g(\pi_\alpha) = x_\alpha^\top \beta, \quad \alpha = 1, \ldots, N, \tag{6.1}$$

where $x_\alpha = (x_{\alpha 1}, \ldots, x_{\alpha p})^\top, \alpha = 1, \ldots, N$ are covariate vectors and $\beta = (\beta_1, \ldots, \beta_p)^\top$ is an unknown parameter vector and $p < N$. When g is a canonical link function, that is,

$$g(t) = \ln\left(\frac{t}{1-t}\right), \tag{6.2}$$

This chapter is based on the following article: Journal of Multivariate Analysis, 123, Taneichi N., Sekiya Y. and Toyama J., Transformed goodness-of-fit statistics for a generalized linear model of binary data, 311–329, Copyright Elsevier (2014).

N. Taneichi and Y. Sekiya, *Improving Tests for Discrete Small Sample Data*, JSS Research Series in Statistics,
https://doi.org/10.1007/978-981-95-5301-3_6

model (6.1) is a logistic regression model. When

$$g(t) = g_P(t) = \Phi^{-1}(t), \tag{6.3}$$

where Φ is the cumulative distribution function of a standard normal distribution, model (6.1) is a probit model. When

$$g(t) = \ln\{-\ln(1-t)\}, \tag{6.4}$$

model (6.1) is a complementary log-log model. Aranda-Ordaz (1981) considered a family of link functions,

$$g(t) = g_c(t) = \ln\left\{\frac{(1-t)^{-c}-1}{c}\right\} \quad (c \geq 0), \tag{6.5}$$

that depend on parameter c. By this family of link functions, we obtain a family of models that includes a logistic regression model when $c = 1$ and a complementary log-log model when $c = 0$ in the limit.

We consider the null hypothesis

$$H_0^g : \pi_\alpha^g = \pi_\alpha(\beta) = g^{-1}(x_\alpha^\top \beta), \quad \alpha = 1, \ldots, N. \tag{6.6}$$

Here, we assume that nuisance parameters of (6.6) are only β. That is, when we use g_c of (6.5) as a link function, we consider c to be fixed. In order to test the null hypothesis H_0^g, we consider the family of ϕ-divergence statistics (Pardo 2006)

$$C_\phi^g = 2\sum_{\alpha=1}^N n_\alpha \left\{\hat{\pi}_\alpha^g \phi\left(\frac{\frac{Y_\alpha}{n_\alpha}}{\hat{\pi}_\alpha^g}\right) + (1-\hat{\pi}_\alpha^g)\phi\left(\frac{1-\frac{Y_\alpha}{n_\alpha}}{1-\hat{\pi}_\alpha^g}\right)\right\}, \tag{6.7}$$

where $\hat{\pi}_\alpha^g = \pi_\alpha(\hat{\beta}^g)$, $\alpha = 1, \ldots, N$, $\hat{\beta}^g = (\hat{\beta}_1^g, \ldots, \hat{\beta}_p^g)^\top$ is the maximum likelihood estimator of β under H_0^g given by (6.6) and ϕ is a real convex function in $(0, \infty)$, satisfying $\phi(1) = \phi'(1) = 0$ and $\phi''(1) = 1$. Here, we note that test statistics C_ϕ^g vary according to link function g. When we choose a convex function ϕ^a given by (2.3) as ϕ, $C_{\phi^a}^g$ becomes a power divergence statistic (Cressie and Read 1984)

$$R_g^a \equiv C_{\phi^a}^g = 2\sum_{\alpha=1}^N n_\alpha \left\{I^a\left(\frac{Y_\alpha}{n_\alpha}, \hat{\pi}_\alpha^g\right) + I^a\left(1-\frac{Y_\alpha}{n_\alpha}, 1-\hat{\pi}_\alpha^g\right)\right\}, \tag{6.8}$$

where $I^a(e, f)$ is given by (3.4).

Under H_0^g, all members of the class of statistics C_ϕ^g defined by (6.7) have a χ_{N-p}^2 limiting distribution, assuming the condition that

$$n_\alpha/n \to \mu_\alpha, \ \alpha = 1, \ldots, N \quad \text{as} \quad n \to \infty, \tag{6.9}$$

where $n = \sum_{\alpha=1}^{N} n_\alpha$, $0 < \mu_\alpha < 1$, $\alpha = 1, \ldots, N$, and $\sum_{\alpha=1}^{N} \mu_\alpha = 1$. Using the results, we can use C_ϕ^g as a goodness-of-fit test statistic for model (6.1). Statistic R_g^0 (the log likelihood ratio statistic or deviance) and statistic R_g^1 (Pearson's X^2 statistic) are used frequently.

6.2 Local Edgeworth Expansion

We consider a local Edgeworth approximation for the probability of Y_α, $\alpha = 1, \ldots, N$ under null hypothesis H_0^g given by (6.6). Let Y_α, $\alpha = 1, \ldots, N$ be distributed according to a binomial distribution $B(n_\alpha, \pi_\alpha^g)$, $\alpha = 1, \ldots, N$, where each π_α^g, $\alpha = 1, \ldots, N$ is represented as $\pi_\alpha^g = g^{-1}(x_\alpha^\top \beta)$, $\alpha = 1, \ldots, N$ by using covariate vectors $x_\alpha = (x_{\alpha 1}, \ldots, x_{\alpha p})^\top$, and an unknown parameter vector β. Let

$$W_\alpha = n_\alpha^{-1/2}(Y_\alpha - n_\alpha \pi_\alpha^g), \quad \alpha = 1, \ldots, N. \tag{6.10}$$

Then, $W = (W_1, \ldots, W_N)^\top$ is a lattice random vector that takes values in the set

$$L = \Big\{ w = (w_1, \ldots, w_N)^\top : w_\alpha = n_\alpha^{-1/2}(y_\alpha - n_\alpha \pi_\alpha^g),\ \alpha = 1, \ldots, N, \\ y = (y_1, \ldots, y_N)^\top \in M \Big\},$$

where

$$M = \Big\{ y = (y_1, \ldots, y_N)^\top : y_1, \ldots, y_N \text{ are non-negative integers that satisfy} \\ y_\alpha \le n_\alpha,\ \alpha = 1, \ldots, N \Big\}.$$

If we consider only for a limiting distribution of C_ϕ^g, we can discuss under the assumption given by (6.9). In this section, since we consider an asymptotic expansion of the distribution of C_ϕ^g, we need an assumption that provides a way of converging n_α/n to μ_α more strictly than that of the assumption given by (6.9). Therefore, we consider the following Assumption 6.1 instead of the assumption given by (6.9).

Assumption 6.1 $n_\alpha \to \infty$, $\alpha = 1, \ldots, N$, as $n \to \infty$, with n_α depending on n in such a way that $n_\alpha/n = \mu_\alpha$, $\alpha = 1, \ldots, N$, where $0 < \mu_\alpha < 1, \alpha = 1, \ldots, N$ and $\sum_{\alpha=1}^{N} \mu_\alpha = 1$.

With regard to a local Edgeworth approximation for the probability of Y_α, $\alpha = 1, \ldots, N$ under H_0^g, we obtain the following theorem.

Theorem 6.1 *For each* $y = (y_1, \ldots, y_N)^\top \in M$, *let* $w = (w_1, \ldots, w_N)^\top$, *where* $w_\alpha = n_\alpha^{-1/2}(y_\alpha - n_\alpha \pi_\alpha^g)$, $\alpha = 1, \ldots, N$. *Then, under Assumption 6.1,*

$$\Pr\{W = w|H_0^g\} = \left(\prod_{\alpha=1}^N n_\alpha^{-1/2}\right) h^g(w)\left\{1 + n^{-1/2}h_1^g(w) + n^{-1}h_2^g(w) + n^{-3/2}h_3^g(w) + O(n^{-2})\right\},$$

where

$$h^g(w) = (2\pi)^{-N/2}|\Omega|^{-1/2}\exp\left(-\frac{1}{2}w^\top\Omega^{-1}w\right), \tag{6.11}$$

$$h_1^g(w) = -\frac{1}{2}\sum_{\alpha=1}^N \frac{1}{\sqrt{\mu_\alpha}}\frac{(1-2\pi_\alpha^g)}{\pi_\alpha^g(1-\pi_\alpha^g)}w_\alpha + \frac{1}{6}\sum_{\alpha=1}^N \frac{1}{\sqrt{\mu_\alpha}}\frac{(1-2\pi_\alpha^g)}{(\pi_\alpha^g)^2(1-\pi_\alpha^g)^2}w_\alpha^3,$$

$$h_2^g(w) = \frac{1}{2}\{h_1^g(w)\}^2 - \frac{1}{12}\sum_{\alpha=1}^N \frac{1}{\mu_\alpha}\frac{(1-\pi_\alpha^g+(\pi_\alpha^g)^2)}{\pi_\alpha^g(1-\pi_\alpha^g)} + \frac{1}{4}\sum_{\alpha=1}^N \frac{1}{\mu_\alpha}\frac{(1-2\pi_\alpha^g+2(\pi_\alpha^g)^2)}{(\pi_\alpha^g)^2(1-\pi_\alpha^g)^2}w_\alpha^2 - \frac{1}{12}\sum_{\alpha=1}^N \frac{1}{\mu_\alpha}\frac{(1-3\pi_\alpha^g+3(\pi_\alpha^g)^2)}{(\pi_\alpha^g)^3(1-\pi_\alpha^g)^3}w_\alpha^4,$$

$$h_3^g(w) = -\frac{1}{3}\{h_1^g(w)\}^3 + h_1^g(w)h_2^g(w) + \frac{1}{12}\sum_{\alpha=1}^N \frac{1}{\mu_\alpha\sqrt{\mu_\alpha}}\frac{(1-2\pi_\alpha)}{(\pi_\alpha^g)^2(1-\pi_\alpha^g)^2}w_\alpha - \frac{1}{6}\sum_{\alpha=1}^N \frac{1}{\mu_\alpha\sqrt{\mu_\alpha}}\frac{(1-2\pi_\alpha^g)(1-\pi_\alpha^g+(\pi_\alpha^g)^2)}{(\pi_\alpha^g)^3(1-\pi_\alpha^g)^3}w_\alpha^3 + \frac{1}{20}\sum_{\alpha=1}^N \frac{1}{\mu_\alpha\sqrt{\mu_\alpha}}\frac{(1-2\pi_\alpha^g)(1-2\pi_\alpha^g+2(\pi_\alpha^g)^2)}{(\pi_\alpha^g)^4(1-\pi_\alpha^g)^4}w_\alpha^5,$$

and

$$\Omega = \mathrm{diag}(\pi_1^g(1-\pi_1^g), \ldots, \pi_N^g(1-\pi_N^g)). \tag{6.12}$$

The proof of Theorem 6.1 is similar to that for Lemma 1 in Taneichi et al. (2011), which is proven by considering the proof of Theorem 22.1 of Bhattacharya and Ranga Rao (1976, pp. 232–236).

6.3 An Approximation for C_ϕ^g Under H_0^g Based on an Asymptotic Expansion

We derive an approximation based on an asymptotic expansion for the distribution of C_ϕ^g under H_0^g. Similar to (3.27), we consider the following approximation for the distribution of C_ϕ^g under H_0^g.

$$\Pr\{C_\phi^g \le x|H_0^g\} \approx J_1^{g,\phi}(x),$$

where $J_1^{g,\phi}(x)$ comes from a multivariate Edgeworth expansion assuming a continuous distribution. With regard to evaluation of $J_1^{g,\phi}(x)$ we obtain the following theorem.

Theorem 6.2 *When g^{-1} is a four times continuously differentiable function and ϕ is a six times continuously differentiable function, under Assumption 6.1, the $J_1^{g,\phi}(x)$ term is evaluated as*

$$J_1^{g,\phi}(x) = \Pr\{\chi^2_{N-p} \le x\} + n^{-1}\sum_{j=0}^{3} v_j^{g,\phi} \Pr\{\chi^2_{N-p+2j} \le x\} + O(n^{-2}), \tag{6.13}$$

where χ_f^2 denotes a chi-square random variable with degrees of freedom f,

$$v_0^{g,\phi} = \frac{1}{24}(-\Gamma_4), \tag{6.14}$$

$$v_1^{g,\phi} = \frac{1}{24}\left[\Gamma_1\phi^{(4)}(1) + \Gamma_2\{\phi'''(1)+1\}^2 + (2\Gamma_1+\Gamma_3)\phi'''(1) + (\Gamma_3+\Gamma_4)\right],$$

$$v_2^{g,\phi} = \frac{1}{24}\left[-\Gamma_1\phi^{(4)}(1) - 2\Gamma_2\{\phi'''(1)+1\}^2 - (2\Gamma_1+\Gamma_3)\phi'''(1) - \Gamma_3\right],$$

$$v_3^{g,\phi} = \frac{1}{24}\Gamma_2\{\phi'''(1)+1\}^2,$$

where

$$\Gamma_1 = -3(A_1 - 2A_3 + A_6),\quad \Gamma_2 = 5A_2 - 12A_4 + 9A_7 - 3B_1 + 6B_2 - 2B_4 - 3B_7,$$

$$\Gamma_3 = 2(3A_1 - 2A_2 - 6A_3 + 6A_4 + 3A_5 + 3A_6 - 6A_7 - 3A_8 - 3B_3 + 2B_4 + 3B_8),$$

$$\Gamma_4 = 6A_1 - 4A_2 - 6A_6 + 12A_8 - 3A_9 + 4B_4 - 12B_5 + 6B_6 - 3B_9,$$

$$A_1 = \sum_{\alpha=1}^{N}\frac{1-3\pi_\alpha^g+3(\pi_\alpha^g)^2}{\mu_\alpha\pi_\alpha^g(1-\pi_\alpha^g)},\quad A_2 = \sum_{\alpha=1}^{N}\frac{(1-2\pi_\alpha^g)^2}{\mu_\alpha\pi_\alpha^g(1-\pi_\alpha^g)},$$

$$A_3 = \sum_{\alpha=1}^{N}\frac{1-3\pi_\alpha^g+3(\pi_\alpha^g)^2}{(\pi_\alpha^g)^2(1-\pi_\alpha^g)^2}G_1(\alpha)^2\sigma_{\alpha\alpha},\quad A_4 = \sum_{\alpha=1}^{N}\frac{(1-2\pi_\alpha^g)^2}{(\pi_\alpha^g)^2(1-\pi_\alpha^g)^2}G_1(\alpha)^2\sigma_{\alpha\alpha},$$

$$A_5 = \sum_{\alpha=1}^{N}\frac{(1-2\pi_\alpha^g)}{\pi_\alpha^g(1-\pi_\alpha^g)}G_2(\alpha)\sigma_{\alpha\alpha},\quad A_6 = \sum_{\alpha=1}^{N}\frac{\mu_\alpha(1-3\pi_\alpha^g+3(\pi_\alpha^g)^2)}{(\pi_\alpha^g)^3(1-\pi_\alpha^g)^3}G_1(\alpha)^4\sigma_{\alpha\alpha}^2,$$

$$A_7 = \sum_{\alpha=1}^{N} \frac{\mu_\alpha(1-2\pi_\alpha^g)^2}{(\pi_\alpha^g)^3(1-\pi_\alpha^g)^3} G_1(\alpha)^4 \sigma_{\alpha\alpha}^2,$$

$$A_8 = \sum_{\alpha=1}^{N} \frac{\mu_\alpha(1-2\pi_\alpha^g)}{(\pi_\alpha^g)^2(1-\pi_\alpha^g)^2} G_1(\alpha)^2 G_2(\alpha) \sigma_{\alpha\alpha}^2,$$

$$A_9 = \sum_{\alpha=1}^{N} \frac{\mu_\alpha}{\pi_\alpha^g(1-\pi_\alpha^g)} G_2(\alpha)^2 \sigma_{\alpha\alpha}^2,$$

$$B_1 = \sum_{\alpha=1}^{N}\sum_{\gamma=1}^{N} \frac{(1-2\pi_\alpha^g)}{\pi_\alpha^g(1-\pi_\alpha^g)} \frac{(1-2\pi_\gamma^g)}{\pi_\gamma^g(1-\pi_\gamma^g)} G_1(\alpha) G_1(\gamma) \sigma_{\alpha\gamma},$$

$$B_2 = \sum_{\alpha=1}^{N}\sum_{\gamma=1}^{N} \frac{\mu_\alpha(1-2\pi_\alpha^g)}{(\pi_\alpha^g)^2(1-\pi_\alpha^g)^2} \frac{(1-2\pi_\gamma^g)}{\pi_\gamma^g(1-\pi_\gamma^g)} G_1(\alpha)^3 G_1(\gamma) \sigma_{\alpha\alpha}\sigma_{\alpha\gamma},$$

$$B_3 = \sum_{\alpha=1}^{N}\sum_{\gamma=1}^{N} \frac{\mu_\alpha}{\pi_\alpha^g(1-\pi_\alpha^g)} \frac{(1-2\pi_\gamma^g)}{\pi_\gamma^g(1-\pi_\gamma^g)} G_1(\alpha) G_2(\alpha) G_1(\gamma) \sigma_{\alpha\alpha}\sigma_{\alpha\gamma},$$

$$B_4 = \sum_{\alpha=1}^{N}\sum_{\gamma=1}^{N} \frac{\mu_\alpha(1-2\pi_\alpha^g)}{(\pi_\alpha^g)^2(1-\pi_\alpha^g)^2} \frac{\mu_\gamma(1-2\pi_\gamma^g)}{(\pi_\gamma^g)^2(1-\pi_\gamma^g)^2} G_1(\alpha)^3 G_1(\gamma)^3 \sigma_{\alpha\gamma}^3,$$

$$B_5 = \sum_{\alpha=1}^{N}\sum_{\gamma=1}^{N} \frac{\mu_\alpha}{\pi_\alpha^g(1-\pi_\alpha^g)} \frac{\mu_\gamma(1-2\pi_\gamma^g)}{(\pi_\gamma^g)^2(1-\pi_\gamma^g)^2} G_1(\alpha) G_2(\alpha) G_1(\gamma)^3 \sigma_{\alpha\gamma}^3,$$

$$B_6 = \sum_{\alpha=1}^{N}\sum_{\gamma=1}^{N} \frac{\mu_\alpha}{\pi_\alpha^g(1-\pi_\alpha^g)} \frac{\mu_\gamma}{\pi_\gamma^g(1-\pi_\gamma^g)} G_1(\alpha) G_2(\alpha) G_1(\gamma) G_2(\gamma) \sigma_{\alpha\gamma}^3,$$

$$B_7 = \sum_{\alpha=1}^{N}\sum_{\gamma=1}^{N} \frac{\mu_\alpha(1-2\pi_\alpha^g)}{(\pi_\alpha^g)^2(1-\pi_\alpha^g)^2} \frac{\mu_\gamma(1-2\pi_\gamma^g)}{(\pi_\gamma^g)^2(1-\pi_\gamma^g)^2} G_1(\alpha)^3 G_1(\gamma)^3 \sigma_{\alpha\alpha}\sigma_{\alpha\gamma}\sigma_{\gamma\gamma},$$

$$B_8 = \sum_{\alpha=1}^{N}\sum_{\gamma=1}^{N} \frac{\mu_\alpha}{\pi_\alpha^g(1-\pi_\alpha^g)} \frac{\mu_\gamma(1-2\pi_\gamma^g)}{(\pi_\gamma^g)^2(1-\pi_\gamma^g)^2} G_1(\alpha) G_2(\alpha) G_1(\gamma)^3 \sigma_{\alpha\alpha}\sigma_{\alpha\gamma}\sigma_{\gamma\gamma},$$

$$B_9 = \sum_{\alpha=1}^{N}\sum_{\gamma=1}^{N} \frac{\mu_\alpha}{\pi_\alpha^g(1-\pi_\alpha^g)} \frac{\mu_\gamma}{\pi_\gamma^g(1-\pi_\gamma^g)} G_1(\alpha) G_2(\alpha) G_1(\gamma) G_2(\gamma) \sigma_{\alpha\alpha}\sigma_{\alpha\gamma}\sigma_{\gamma\gamma},$$

$$G_i(\alpha) = u^{(i)}(x_\alpha^\top \beta), \quad \alpha = 1, \ldots, N, \ i = 1, 2,$$

$$u(x) = g^{-1}(x),$$

$$\sigma_{\alpha\gamma} = \sum_{\ell=1}^{p} \sum_{m=1}^{p} \kappa^{\ell,m} x_{\alpha\ell} x_{\gamma m}, \qquad \alpha, \gamma = 1, \ldots, N,$$

$$\kappa_{\ell,m} = \sum_{\lambda=1}^{N} \mu_\lambda \{\pi_\lambda^g (1 - \pi_\lambda^g)\}^{-1} G_1(\lambda)^2 x_{\lambda\ell} x_{\lambda m}, \qquad \ell, m = 1, \ldots, p,$$

where $u^{(i)}$ is the ith derivative of u and $\kappa^{\ell,m}$ is the (ℓ, m)- element of the inverse matrix K^{-1} of $K = (\kappa_{\ell,m})$.

Proof of Theorem 6.2 is shown in Appendix ??. We find that the coefficients $v_j^{g,\phi}$, $j = 0, 1, 2, 3$ satisfy the relation $\sum_{j=0}^{3} v_j^{g,\phi} = 0$. If we apply ϕ^a as ϕ in Theorem 6.2, we obtain the following corollary for the statistic R_g^a based on power divergence.

Corollary 6.1 *When the statistic is R_g^a given by (6.8) and g^{-1} is a fourth time continuously differentiable function, under Assumption 6.1, the $J_1^{g,\phi}(x)$ term is evaluated as*

$$J_1^{g,\phi}(x) = \Pr\{\chi_{N-p}^2 \leq x\} + n^{-1} \sum_{j=0}^{3} v_j^{g,(a)} \Pr\{\chi_{N-p+2j}^2 \leq x\} + O(n^{-2}), \qquad (6.15)$$

where $v_j^{g,(a)}$, $j = 0, 1, 2, 3$ are defined as $v_j^{g,\phi}$, $j = 0, 1, 2, 3$ in the case of $\phi'''(1) = a - 1$ and $\phi^{(4)}(1) = (a - 1)(a - 2)$, respectively.

We give some examples with regard to Theorem 6.2. When link function g is given by (6.2), that is, the logistic regression model, then

$$g^{-1}(x) = \frac{\exp(x)}{1 + \exp(x)},$$

$$G_1(\alpha) = \pi_\alpha^g (1 - \pi_\alpha^g), \quad \alpha = 1, \ldots, N,$$

and

$$G_2(\alpha) = \pi_\alpha^g (1 - \pi_\alpha^g)(1 - 2\pi_\alpha^g), \quad \alpha = 1, \ldots, N$$

in Theorem 6.2 Applying the above results to Corollary 6.1 in the case of $a = 0$, the expansion (6.15) coincides with expansion (2.5) in Taneichi et al. (2011), that is, an expansion for the deviance for the logistic regression model. When link function g is given by (6.3), that is, the probit model, then

$$g^{-1}(x) = \Phi(x),$$

$$G_1(\alpha) = \frac{1}{\sqrt{2\pi}} \exp\left[-\frac{\{\Phi^{-1}(\pi_\alpha^g)\}^2}{2} \right], \quad \alpha = 1, \ldots, N,$$

and

$$G_2(\alpha) = -\frac{1}{\sqrt{2\pi}} \Phi^{-1}(\pi_\alpha^g) \exp\left[-\frac{\{\Phi^{-1}(\pi_\alpha^g)\}^2}{2} \right], \quad \alpha = 1, \ldots, N$$

in Theorem 6.2. When link function g is given by (6.4), that is, the complementary log-log model, then

$$g^{-1}(x) = 1 - \exp\{-\exp(x)\},$$

$$G_1(\alpha) = -(1 - \pi_\alpha^g) \ln(1 - \pi_\alpha^g), \quad \alpha = 1, \ldots, N,$$

and

$$G_2(\alpha) = -(1 - \pi_\alpha^g)\{\ln(1 - \pi_\alpha^g)\}\{1 + \ln(1 - \pi_\alpha^g)\}, \quad \alpha = 1, \ldots, N$$

in Theorem 6.2. When link function g is given by (6.5), that is, the family of models proposed by Aranda-Ordaz (1981), then

$$g^{-1}(x) = 1 - \{1 + c \exp(x)\}^{-1/c} \ (c \geq 0),$$

$$G_1(\alpha) = c^{-1}(1 - \pi_\alpha^g)\{1 - (1 - \pi_\alpha^g)^c\} \ (c \geq 0), \quad \alpha = 1, \ldots, N,$$

and

$$G_2(\alpha) = c^{-2}(1 - \pi_\alpha^g)\{1 - (1 - \pi_\alpha^g)^c\}\{(c+1)(1 - \pi_\alpha^g)^c - 1\} \ (c \geq 0), \ \alpha = 1, \ldots, N$$

in Theorem 6.2

6.4 A Transformed Test Statistic for a Generalized Linear Model of Binary Data

Applying the transformed statistics T_B given by (2.10), T_I given by (2.11), and T_{CF} given by (2.12) to the evaluation (6.13) given by Theorem 6.2, we construct transformations for improving small sample accuracy of the chi-square approximation of the distribution of C_ϕ^g under H_0^g.

When ϕ satisfies (4.9), equations $v_1^{g,\phi} = -v_0^{g,\phi}$ and $v_2^{g,\phi} = v_3^{g,\phi} = 0$ hold in Theorem 6.2. Then, we can consider Bartlett adjustment

$$C^B_{g,\phi} = \left\{1 + \frac{2v_0^{g,\phi}}{n(N-p)}\right\} C^g_{\phi}.$$

On the other hand, when ϕ does not satisfy (4.9), we can consider the transformed statistic (Improved transformation)

$$C^I_{g,\phi} = (n\alpha + \beta)^2 \ln(1+\zeta),$$

where

$$\zeta = \frac{1}{(n\alpha)^2}\left[C^g_{\phi} + \frac{1}{n\alpha}\{(C^g_{\phi})^2 + \gamma (C^g_{\phi})^3\} + \frac{1}{(n\alpha)^2}\left\{\frac{1}{3}(C^g_{\phi})^3 + \frac{3\gamma}{4}(C^g_{\phi})^4 + \frac{9\gamma^2}{20}(C^g_{\phi})^5\right\}\right],$$

$\alpha = -(N-p)(N-p+2)\{2(v_2^{g,\phi} + v_3^{g,\phi})\}^{-1}$, $\beta = -(N-p+2)v_0^{g,\phi}\{2(v_2^{g,\phi} + v_3^{g,\phi})\}^{-1}$ and $\gamma = v_3^{g,\phi}\{(N-p+4)(v_2^{g,\phi} + v_3^{g,\phi})\}^{-1}$.

We can also consider transformed statistic (Bartlett-type adjustment)

$$C^{CF}_{g,\phi} = \left\{1 + \frac{2v_0^{g,\phi}}{n(N-p)} + \frac{2(v_0^{g,\phi} + v_1^{g,\phi})}{N(N-p)(N-p+2)} C^g_{\phi} + \frac{2(v_0^{g,\phi} + v_1^{g,\phi} + v_2^{g,\phi})}{N(N-p)(N-p+2)(N-p+4)} (C^g_{\phi})^2\right\} C^g_{\phi}.$$

In practical application, we may use estimate $\hat{v}_j^{g,\phi}$, $j = 0, 1, 2, 3$ obtained by substituting maximum likelihood estimate $\hat{\beta}^g$ for true value β in $v_j^{g,\phi}$, $j = 0, 1, 2, 3$.

In the case of power divergence statistic $R^a_g \equiv C^g_{\phi^a}$, condition (4.9) is satisfied if and only if $a = 0$ (the log likelihood ratio statistic). Then, we consider the transformed statistic C^B_{g,ϕ^0} when $a = 0$ and put $R^0_B \equiv C^B_{g,\phi^0}$. When the link function g is defined by (6.2), that is, a logistic regression model, statistic R^0_B coincides with the statistic D^* proposed by (3.3) in Taneichi et al. (2011). On the other hand, we consider statistic C^I_{g,ϕ^a} and C^{CF}_{g,ϕ^a} when $a \neq 0$ and put R^a_I and R^a_{CF}, respectively. We note that $R^0_{CF} = R^0_B$, then R^a_{CF} can be defined when $a = 0$ as well as when $a \neq 0$.

Taneichi et al. (2014) investigated the small sample performance of approximation by a chi-square distribution of goodness-of-fit test statistic for a generalized linear model of binary data. For power divergence statistics, performance of transformed statistics R^0_B and R^a_I $(a = 0.2, 2/3, 1)$ were numerically compared with that of the original statistic R^a_g $(a = 0, 0.2, 2/3, 1)$ for significance levels 0.01, 0.05, and 0.1. For concrete models, the following 4 models were considered. The Aranda-Ordaz model with $c = 0$ (complementary log-log model), with $c = 1/2$, and with $c = 1$ (logistic regression model), and a probit model. The results of the comparisons were as follows. Transformed statistics R^0_B and $R^{0.2}_I$ performed better than the original statistics R^0_g and $R^{0.2}_g$ for all models. R^a_I $(a = 2/3, 1)$ performed better than the original statistics R^a_g $(a = 2/3, 1)$ for 3 models including the logistic regression model and the probit model. However, R_I^a $(a = 2/3, 1)$ did not perform better than the original statistics R_g^a $(a = 2/3, 1)$ in some situations for the complementary log-log model. In Taneichi et al. (2014), the power against an alternative hypothesis of transformed statistics R_B^0, and R_I^a $(a = 0.2, 2/3, 1)$ were compared with that of the original statistics R_g^a $(a = 0, 0.2, 2/3, 1)$. The results of comparisons were as follows. The power of transformed statistics R_B^0 and R_I^a $(a = 0.2, 2/3, 1)$ were almost the same as that of the original statistics R_g^a $(a = 0, 0.2, 2/3, 1)$ for the 4 considered models.

Furthermore, Taneichi et al. (2016) derived transformed goodness-of-fit statistics for a generalized linear model of binary data when the parameter is not estimated by maximum likelihood estimation but by estimated by minimum ϕ-divergence estimation.

References

Aranda-Ordaz FJ (1981) On two families of transformations to additivity for binary response data. Biometrika 68:357–363

Bhattacharya RN, Ranga Rao R (1976) Normal approximation and asymptotic expansions. Wiley, New York

Cressie N, Read TRC (1984) Multinomial goodness-of-fit tests. J Roy Statist Soc Ser B 46:440–464

Nelder JA, Wedderburn RWM (1972) Generalized linear models. J R Statist Soc A 135:370–384

Pardo L (2006) Statistical inference based on divergence measures. Chapman & Hall/CRC

Taneichi N, Sekiya Y, Toyama J (2011) Improved transformed deviance statistic for testing a logistic regression model. J Multivariate Anal 102:1263–1279

Taneichi N, Sekiya Y, Toyama J (2014) Transformed goodness-of-fit statistics for a generalized linear model of binary data. J Multivariate Anal 123:311–329

Taneichi N, Sekiya Y, Toyama J (2016) Improved transformation of $\phi-$divergence goodness-of-fit test statistics based on minimum ϕ^*-divergence estimator for GLIM of binary data. SUT J Math 52(2):193–215

Chapter 7
A Transformed Goodness-of-Fit Statistic for a Loglinear Model in a Contingency Table

Keywords Asymptotic expansion · Contingency table · Loglinear model · LR statistic · Transformed statistics

7.1 A Loglinear Model for a Contingency Table and Its Goodness-of-Fit Test Statistic

We can systematically treat a variety of independence models of a multi-way contingency table by representation as a loglinear model including interaction terms. That is, since parameters of a loglinear model correspond to main effect terms and interaction terms, respectively, each independence model is represented as a model in which some of the parameters for interaction terms of the full model are zero. Furthermore, if we are not particular about the restriction of a hierarchical model, we can systematically treat models of symmetry as well as models of independence.

In this chapter, we consider a loglinear model for multinomial sampling (e.g., Agresti, 2013). We can consider a M-way $J_1 \times \cdots \times J_M$ table as N-dimensional data $y = (y_1, \ldots, y_N)^\top$ by renumbering the index of each element, where $N = \prod_{m=1}^{M} J_m$ is the total number of cells. Let $Y = (Y_1, \ldots, Y_N)^\top$ be distributed as a multinomial distribution $\mathcal{M}_N(n, \pi)$, where $\pi = (\pi_1, \ldots, \pi_N)^\top, 0 < \pi_\alpha < 1,\ \alpha = 1, \ldots, N$ and $\sum_{\alpha=1}^{N} \pi_\alpha = 1$. A loglinear model for multinomial sampling is represented as

$$\pi = \pi(\theta) \equiv \frac{\exp(X\theta)}{1_N^\top \exp(X\theta)}, \tag{7.1}$$

where $X = (X_1, \ldots, X_N)^\top = (x_{\alpha j})$ is an $N \times p$ design matrix or (model matrix), $\theta = (\theta_1, \ldots, \theta_p)^\top$ $(p < N)$ is an unknown parameter vector, 1_N is an N-dimensional

N. Taneichi and Y. Sekiya, *Improving Tests for Discrete Small Sample Data*,
JSS Research Series in Statistics,
https://doi.org/10.1007/978-981-95-5301-3_7

column vector with unit elements, $\pi(\theta) = (\pi_1(\theta), \dots, \pi_N(\theta))^\top$ and $\exp(a)$ is defined by $(e^{a_1}, \dots, e^{a_N})^\top$ when $a = (a_1, \dots, a_N)^\top$. In order to test the null hypothesis

$$H_0^L : \text{Model given by (7.1) is true},$$

we consider the log likelihood ratio test statistic

$$G^2 = 2\sum_{\alpha=1}^{N} Y_\alpha \ln\left(\frac{Y_\alpha}{n\pi_\alpha(\hat{\theta})}\right), \tag{7.2}$$

where $\pi_\alpha(\hat{\theta}), \alpha = 1, \dots, N$ are defined by $\pi_\alpha(\hat{\theta}) = \exp(X_\alpha^\top\hat{\theta})/\{1_N^\top \exp(X\hat{\theta})\}$, where $\hat{\theta}$ is MLE of θ under H_0^L. Under the null hypothesis H_0^L, it is known that the log likelihood ratio test statistic G^2 has a chi-square limiting distribution with degrees of freedom $N - p - 1$. By using the results for a large sample, we can use the log likelihood ratio test statistic for a goodness-of-fit test of a loglinear model (7.1).

7.2 Examples of Loglinear Models for a Contingency Table

In this section, we show examples of loglinear models for a contingency table. The examples correspond to various independence models for a contingency table.

When we consider a 3-way 2×2×2 contingency table, by using the notation of Agresti (2013), a full loglinear model for Poisson sampling is given by

$$\ln m_{jk\ell} = \mu + \lambda_j^X + \lambda_k^Y + \lambda_\ell^Z + \lambda_{jk}^{XY} + \lambda_{j\ell}^{XZ} + \lambda_{k\ell}^{YZ} + \lambda_{jk\ell}^{XYZ},\ j = 1, 2, k = 1, 2, \ell = 1, 2,$$

where $m_{jk\ell}$ $(j = 1, 2, k = 1, 2, \ell = 1, 2)$ is the expectation of each cell. By considering the constraints $\sum_{j=1}^2 \lambda_j^X = \sum_{k=1}^2 \lambda_k^Y = \sum_{\ell=1}^2 \lambda_\ell^Z = 0$, $\sum_{j=1}^2 \lambda_{jk}^{XY} = \sum_{k=1}^2 \lambda_{jk}^{XY} = \sum_{j=1}^2 \lambda_{j\ell}^{XZ} = \sum_{\ell=1}^2 \lambda_{j\ell}^{XZ} = \sum_{k=1}^2 \lambda_{k\ell}^{YZ} = \sum_{\ell=1}^2 \lambda_{k\ell}^{YZ} = 0$ and $\sum_{j=1}^2 \lambda_{jk\ell}^{XYZ} = \sum_{k=1}^2 \lambda_{jk\ell}^{XYZ} = \sum_{\ell=1}^2 \lambda_{jk\ell}^{XYZ} = 0$, we rewrite the model using vectors and a matrix,

$$\begin{aligned}&(\ln m_{111}, \ln m_{112}, \ln m_{121}, \ln m_{122}, \ln m_{211}, \ln m_{212}, \ln m_{221}, \ln m_{222})^\top \\ &= (1_8 \vdots X)(\mu \vdots \theta^\top)^\top,\end{aligned}$$

where

$$X = \begin{pmatrix} 1 & 1 & 1 & 1 & 1 & 1 & 1 \\ 1 & 1 & -1 & 1 & -1 & -1 & -1 \\ 1 & -1 & 1 & -1 & 1 & -1 & -1 \\ 1 & -1 & -1 & -1 & -1 & 1 & 1 \\ -1 & 1 & 1 & -1 & -1 & 1 & -1 \\ -1 & 1 & -1 & -1 & 1 & -1 & 1 \\ -1 & -1 & 1 & 1 & -1 & -1 & 1 \\ -1 & -1 & -1 & 1 & 1 & 1 & -1 \end{pmatrix}, \quad \theta = \begin{pmatrix} \lambda_1^X \\ \lambda_1^Y \\ \lambda_1^Z \\ \lambda_{11}^{XY} \\ \lambda_{11}^{XZ} \\ \lambda_{11}^{YZ} \\ \lambda_{111}^{XYZ} \end{pmatrix}, \quad \mu = \ln n - \ln S,$$

$(\,\vdots\,)$ is a partition matrix, $S = 1_8^\top \exp(X\theta)$ and n is the sample size. The corresponding full loglinear model for multinomial sampling in a 2×2×2 contingency table is given by

$$(\pi_{111}, \pi_{112}, \pi_{121}, \pi_{122}, \pi_{211}, \pi_{212}, \pi_{221}, \pi_{222})^\top = \frac{\exp(X\theta)}{1_8^\top \exp(X\theta)}, \tag{7.3}$$

where $\pi_{jk\ell}$ ($j = 1, 2, k = 1, 2, \ell = 1, 2$) is a cell probability. Then, we can consider the following models.

(I) The model when $\lambda_{111}^{XYZ} = 0$, that is, when a three-factor interaction term does not exist. The model is represented as substituting $X'\theta'$ for $X\theta$ in (7.3) when θ' is the vector defined by eliminating the 7th row of θ and X' is the matrix defined by eliminating the 7th column of the design matrix X. The model corresponds to the model for 3-way tables developed by Bartlett (1935) that must be stated in terms of odds ratios in a 2×2×2 contingency table. The log likelihood ratio test statistic G^2 is asymptotically distributed as a chi-square distribution with degrees of freedom 1 under the model.

(II) The model when $\lambda_{111}^{XYZ} = 0$ and $\lambda_{11}^{YZ} = 0$. This model corresponds to that of conditional independence of Y and Z given X in a 2×2×2 contingency table. The log likelihood ratio test statistic G^2 is asymptotically distributed as a chi-square distribution with degrees of freedom 2 under the model.

(III) The model when $\lambda_{111}^{XYZ} = 0$, $\lambda_{11}^{YZ} = 0$ and $\lambda_{11}^{XZ} = 0$. This model corresponds to that for independence of Z and joint distribution of X and Y in a 2×2×2 contingency table. The log likelihood ratio test statistic G^2 is asymptotically distributed as a chi-square distribution with degrees of freedom 3 under the model.

(IV) The model when $\lambda_{111}^{XYZ} = 0$, $\lambda_{11}^{YZ} = 0$, $\lambda_{11}^{XZ} = 0$ and $\lambda_{11}^{XY} = 0$. This model corresponds to that of complete independence in a 2×2×2 contingency table. The log likelihood ratio test statistic G^2 is asymptotically distributed as a chi-square distribution with degrees of freedom 4 under the model.

Similar to a 2×2×2 contingency table, a full loglinear model for multinomial sampling in a 2×2×3 contingency table is given by

$$(\pi_{111}, \pi_{112}, \pi_{113}, \pi_{121}, \pi_{122}, \pi_{123}, \pi_{211}, \pi_{212}, \pi_{213}, \pi_{221}, \pi_{222}, \pi_{223})^\top$$

$$= \frac{\exp(X\theta)}{1_{12}^{\top} \exp(X\theta)}, \tag{7.4}$$

where $\theta = (\lambda_1^X, \lambda_1^Y, \lambda_1^Z, \lambda_2^Z, \lambda_{11}^{XY}, \lambda_{11}^{XZ}, \lambda_{12}^{XZ}, \lambda_{11}^{YZ}, \lambda_{12}^{YZ}, \lambda_{111}^{XYZ}, \lambda_{112}^{XYZ})^{\top}$ and X is a suitable 12×11 design matrix. We can consider model (I) for a $2\times2\times3$ contingency table, which is the model when $\lambda_{111}^{XYZ} = \lambda_{112}^{XYZ} = 0$. The model is represented as substituting $X''\theta''$ for $X\theta$ in (7.4) when θ'' is the vector defined by eliminating the10th and 11th rows of θ and X'' is the matrix defined by eliminating the 10th and 11th columns of the design matrix X. G^2 is asymptotically distributed as a chi-square distribution with degrees of freedom 2 under the model.

The following four independence models for a 3-way contingency table are usually considered. (a) The independence model developed by Bartlett (1935). (b) The conditional independence model. (c) A jointly independence model. (d) The complete independence model. Transformed test statistics for models (b), (c), and (d) were derived in Chaps. 4 and 5. However, transformed test statistics for model (a) has not been derived. The tests of independence for a 3-way contingency table are covered by the loglinear models corresponding to (I), (II), (III), and (IV), even when it is not a $2 \times 2 \times 2$ model. Therefore, by considering the loglinear model corresponding to (I), we can derive transformed statistics for model (a).

7.3 An Approximation for LR Statistic Under H_0^L Based on an Asymptotic Expansion

We derive a local Edgeworth approximation for the probability of Y under the null hypothesis H_0^L. Let

$$U_\alpha = n^{-1/2}(Y_\alpha - n\pi_\alpha(\theta)), \quad \alpha = 1, \ldots, N-1. \tag{7.5}$$

Then, $U = (U_1, \ldots, U_{N-1})^{\top}$ is a lattice random vector that takes values in the set

$$S_1 = \{u = (u_1, \ldots, u_{N-1})^{\top} : u_\alpha = n^{-1/2}(y_\alpha - n\pi_\alpha(\theta)),$$
$$\tilde{y} = (y_1, \ldots, y_{N-1})^{\top} \in S_0\},$$

where $S_0 = \{\tilde{y} = (y_1, \ldots, y_{N-1})^{\top} : y_1, \ldots, y_{N-1}$ are non-negative integers and $\sum_{\alpha=1}^{N-1} y_\alpha \leq n\}$. With regard to a local Edgeworth approximation for the probability of Y under H_0^L, we obtain the following theorem.

Theorem 7.1 *For each $\tilde{y} \in S_0$, let $u = (u_1, \ldots, u_{N-1})^{\top}$, where $u = n^{-1/2}(\tilde{y} - n\tilde{\pi}(\theta))$ and $\tilde{\pi}(\theta) = (\pi_1(\theta), \ldots, \pi_{N-1}(\theta))^{\top}$. Then, the probability that $U = u$ under H_0^L is evaluated as follows:*

$$\Pr\{U = u | H_0^L\} = n^{-(N-1)/2} f(u) \left\{ 1 + n^{-1/2} h_1(u) + n^{-1} h_2(u) + n^{-3/2} h_3(u) + O(n^{-2}) \right\},$$

where

$$f(u) = (2\pi)^{-(N-1)/2} \mid \Omega \mid^{-1/2} \exp\left(-\frac{1}{2} u^\top \Omega^{-1} u\right),$$

$$\Omega = \mathrm{diag}(\pi_1(\theta), \ldots, \pi_{N-1}(\theta)) - \tilde{\pi}(\theta)\tilde{\pi}(\theta)^\top, \tag{7.6}$$

$$h_1(u) = -\frac{1}{2} M_1^1 + \frac{1}{6} M_2^3, \quad h_2(u) = \frac{1}{2}\{h_1(u)\}^2 + \frac{1}{12}(1 - M_1^0) + \frac{1}{4} M_2^2 - \frac{1}{12} M_3^4,$$

$$h_3(u) = -\frac{1}{3}\{h_1(u)\}^3 + h_1(u) h_2(u) + \frac{1}{12} M_2^1 - \frac{1}{6} M_3^3 + \frac{1}{20} M_4^5,$$

and $M_b^a = \sum_{\alpha=1}^N u_\alpha^a / \{\pi_\alpha(\theta)\}^b$.

The proof of Theorem 7.1 is shown in Appendix A.7.

Next, we derive an approximation of the distribution of the log likelihood ratio statistic G^2 under H_0^L based on an asymptotic expansion. Similar to (3.27), we consider the approximation $\Pr\{G^2 \leq x | H_0^L\} \approx J_1^*(x)$, where $J_1^*(x)$ comes from a multivariate Edgeworth expansion. The following result, proved in Appendix A.8, details how the term $J_1^*(x)$ can be evaluated.

Theorem 7.2 *The continuous term $J_1^*(x)$ of asymptotic expansion for the distribution of the log likelihood ratio test statistic G^2 for a loglinear model is evaluated as follows:*

$$J_1^*(x) = \Pr\{\chi_\nu^2 \leq x\} + n^{-1} \sum_{j=0}^{1} d_j \Pr\{\chi_{\nu+2j}^2 \leq x\} + O(n^{-2}), \tag{7.7}$$

where

$$d_0 = \frac{1}{24}\Big\{2 - 2M_1^0 + 6A_1 - 3A_2 + 6A_3 - 8A_4 + 2A_5 + 3A_6 - 4A_7 + 3(A_1)^2 + 6A_1A_3 - 6A_1A_4 + 6(A_3)^2 + 3(A_1)^2 A_3 + 2(A_3)^3\Big\},$$

$$d_1 = -d_0, \quad M_1^0 = \sum_{\alpha=1}^N \pi_\alpha^{-1}, \quad A_1 = \sum_{\alpha=1}^N \pi_\alpha \sigma_{\alpha\alpha}, \quad A_2 = \sum_{\alpha=1}^N \pi_\alpha \sigma_{\alpha\alpha}^2,$$

$$A_3 = \sum_{\alpha=1}^{N}\sum_{\beta=1}^{N}\pi_\alpha\pi_\beta\sigma_{\alpha\beta}, \quad A_4 = \sum_{\alpha=1}^{N}\sum_{\beta=1}^{N}\pi_\alpha\pi_\beta\sigma_{\alpha\alpha}\sigma_{\alpha\beta}, \quad A_5 = \sum_{\alpha=1}^{N}\sum_{\beta=1}^{N}\pi_\alpha\pi_\beta\sigma_{\alpha\beta}^3,$$

$$A_6 = \sum_{\alpha=1}^{N}\sum_{\beta=1}^{N}\pi_\alpha\pi_\beta\sigma_{\alpha\alpha}\sigma_{\alpha\beta}\sigma_{\beta\beta}, \quad A_7 = \sum_{\alpha=1}^{N}\sum_{\beta=1}^{N}\sum_{\gamma=1}^{N}\pi_\alpha\pi_\beta\pi_\gamma\sigma_{\alpha\beta}^2\sigma_{\alpha\gamma},$$

$$\sigma_{\alpha\beta} = \sum_{j=1}^{p}\sum_{k=1}^{p}D^{jk}x_{\alpha j}x_{\beta k}, \quad D_{jk} = \sum_{\gamma=1}^{N}\sum_{\epsilon=1}^{N}(\delta_{\gamma\epsilon}\pi_\gamma - \pi_\gamma\pi_\epsilon)x_{\gamma j}x_{\epsilon k},$$

D^{jk} is the (j, k)- element of the inverse matrix D^{-1} of $D = (D_{jk})$, $x_{\alpha j}$ is the (α, j)- element of the $N \times p$ design matrix X, $\nu = N - p - 1$, N is the total number of cells of the contingency table, p is the dimension of parameter vector θ and $\pi_\alpha = \pi_\alpha(\theta), \alpha = 1, \ldots, N$.

7.4 A Transformed Statistic of the LR Test Statistic Based on a Continuous Term of Asymptotic Expansion

In this section, when the distribution of log likelihood ratio statistic G^2 under H_0 is approximated by $J_1^*(x)$, we construct a transformation of statistic G^2 that improves small sample accuracy of the chi-square approximation. Since $d_1 = -d_0$ holds in Theorem 7.2, applying (2.10) to the evaluation (7.7), we obtain a Bartlett adjustment (7.8) as follows:

$$G_B^2 = \left\{1 + \frac{2d_0}{n(N - p - 1)}\right\} G^2. \tag{7.8}$$

We call this a Bartlett-type transformed statistic.

Now, the coefficient d_0 includes $\pi(\theta)$, which is a function of the unknown parameter vector $\theta = (\theta_1, \ldots, \theta_p)^\top$. Then, in a practical application, we may use estimates $\pi_\alpha(\hat{\theta})$, $\alpha = 1, \ldots, N$ that are obtained by substituting the maximum likelihood estimate $\hat{\theta}$ for θ.

The approximation $\Pr\{G_B^2 < x\} \approx \Pr\{\chi_{N-p-1}^2 < x\}$ is justified by (2.13). On the other hand, this approximation is rewritten as

$$\Pr\{G_B^2 \geq x\} \approx \Pr\{\chi_{N-p-1}^2 \geq x\}. \tag{7.9}$$

(7.9) means that the p-values of the test statistic G_B^2 are approximated well by $\Pr\{\chi_{N-p-1}^2 \geq x\}$.

To increase the speed of convergence to a chi-square limiting distribution, we separately constructed the transformed test statistic G_B^2 of the log likelihood ratio test statistic G^2 for testing some kinds of independence in a contingency table. By considering a loglinear model, we can construct the transformed test statistic G_B^2 of G^2 systematically, as a specific example. Against the full loglinear model

$$(\pi_{111}, \ldots, \pi_{JKL})^\top = \frac{\exp(X\theta)}{1_{JKL}^\top \exp(X\theta)} \tag{7.10}$$

in a 3-way $J \times K \times L$ contingency table, where $\theta = (\lambda_1^X, \ldots, \lambda_{J-1}^X, \lambda_1^Y, \ldots, \lambda_{K-1}^Y, \lambda_1^Z, \ldots, \lambda_{L-1}^Z, \lambda_{1\,1}^{XY}, \ldots, \lambda_{J-1\,K-1}^{X\ \ Y}, \lambda_{1\,1}^{XZ}, \ldots, \lambda_{J-1\,L-1}^{X\ \ Z}, \lambda_{1\,1}^{YZ}, \ldots, \lambda_{K-1\,L-1}^{Y\ \ Z}, \lambda_{1\,1\,1}^{XYZ}, \ldots, \lambda_{J-1\,K-1\,L-1}^{X\ \ Y\ \ Z})^\top$ and X is a suitable $JKL \times (JKL-1)$ design matrix, the transformed test statistic G_B^2 of G^2 for the model $\lambda_{j\,k}^{XY} = \lambda_{j\,\ell}^{XZ} = \lambda_{k\,\ell}^{YZ} = \lambda_{j\,k\,\ell}^{XYZ} = 0$ for all j, k, ℓ corresponds to the transformed test statistic of the complete independence of a 3-way contingency table derived by Taneichi et al. (2021). Against (7.10), the transformed test statistic G_B^2 of G^2 for the model $\lambda_{j\,\ell}^{XZ} = \lambda_{k\,\ell}^{YZ} = \lambda_{j\,k\,\ell}^{XYZ} = 0$ for all j, k, ℓ corresponds to the transformed statistic of two factors independence from another in a 3-way contingency table derived by Kobe et al. (2015). Against (7.10), the transformed test statistic G_B^2 of G^2 for the model $\lambda_{k\,\ell}^{YZ} = \lambda_{j\,k\,\ell}^{XYZ} = 0$ for all j, k, ℓ corresponds to the tansformed test statistic of conditional independence of a 3-way contingency table derived by Taneichi et al. (2019). On the other hand, for the model of 3-way tables developed by Bartlett (1935), which must be expressed in terms of odds ratios, a transformed statistic of the log likelihood ratio statistic has not yet been derived separately. However, we can obtain the transformed test statistic G_B^2 of G^2 by considering the model $\lambda_{j\,k\,\ell}^{XYZ} = 0$ for all j, k, ℓ against (7.10). Furthermore, the transformed statistic of the log likelihood ratio statistic of independence among groups of factors in an M-way contingency table derived by Taneichi et al. (2021) can also be derived as a specific example by considering a loglinear model for an M-way contingency table. In various kinds of loglinear models for an M-way contingency table, we can derive the transformed statistic G_B^2 of log likelihood ratio statistic G^2 that increases the speed of convergence to a chi-square limiting distribution.

7.5 Numerical Comparison of the LR Test Statistic G^2 and the Transformed Statistic G_B^2

In this section, we numerically compare the performance of the Bartlett-type transformed statistic G_B^2 with that of the original log likelihood ratio statistic G^2. We consider a model in which three-factor interaction terms do not exist in 2×2×2 and 2×2×3 contingency tables.

First, in a 2×2×2 contingency table, we consider multinomial sampling model (I) of (7.3). In this model, let the true value of θ be $\theta^* = (\lambda_1^{X*}, \lambda_1^{Y*}, \lambda_1^{Z*}, \lambda_{1\,1}^{XY*}, \lambda_{1\,1}^{XZ*}, \lambda_{1\,1}^{YZ*})^\top$. Then, the true value of π is $\pi^* = (\pi_{111}^*, \pi_{112}^*, \pi_{121}^*, \pi_{122}^*, \pi_{211}^*, \pi_{212}^*, \pi_{221}^*, \pi_{222}^*)^\top$, where π^* is obtained by substituting $X'\theta^*$ for $X\theta$ in (7.3), where X' is defined in Sect. 7.2. We investigated the following four cases as true parameters.

(i) $\lambda_1^{X*} = 0.05,\ \lambda_1^{Y*} = 0.03,\ \lambda_1^{Z*} = 0.02,\ \lambda_{11}^{XY*} = 0.1,\ \lambda_{11}^{XZ*} = 0.2,\ \lambda_{11}^{YZ*} = 0.15$
(ii) $\lambda_1^{X*} = 0.1,\ \lambda_1^{Y*} = 0.1,\ \lambda_1^{Z*} = 0.1,\ \lambda_{11}^{XY*} = 0.2,\ \lambda_{11}^{XZ*} = 0.2,\ \lambda_{11}^{YZ*} = 0.2$
(iii) $\lambda_1^{X*} = 0.1,\ \lambda_1^{Y*} = 0.1,\ \lambda_1^{Z*} = 0.2,\ \lambda_{11}^{XY*} = 0.1,\ \lambda_{11}^{XZ*} = 0.1,\ \lambda_{11}^{YZ*} = 0.2$
(iv) $\lambda_1^{X*} = 0.2,\ \lambda_1^{Y*} = 0.2,\ \lambda_1^{Z*} = 0.2,\ \lambda_{11}^{XY*} = 0.1,\ \lambda_{11}^{XZ*} = 0.1,\ \lambda_{11}^{YZ*} = 0.1$

Next, in a 2×2×3 contingency table, we consider multinomial sampling model (I) of (7.4). In this model, let the true value of θ be $\theta^* = (\lambda_1^{X*}, \lambda_1^{Y*}, \lambda_1^{Z*}, \lambda_2^{Z*}, \lambda_{11}^{XY*}, \lambda_{11}^{XZ*}, \lambda_{12}^{XZ*}, \lambda_{11}^{YZ*}, \lambda_{12}^{YZ*})^\top$. Then, the true value of π is $\pi^* = (\pi_{111}^*, \pi_{112}^*, \pi_{113}^*, \pi_{121}^*, \pi_{122}^*, \pi_{123}^*, \pi_{211}^*, \pi_{212}^*, \pi_{213}^*, \pi_{221}^*, \pi_{222}^*, \pi_{223}^*)^\top$, where π^* is obtained by substituting $X''\theta^*$ for $X\theta$ in (7.4), where X'' is defined in Sect. 7.2. We investigated the following four cases as true parameters.

(v) $\lambda_1^{X*} = 0.05,\ \lambda_1^{Y*} = 0.03,\ \lambda_1^{Z*} = 0.02,\ \lambda_2^{Z*} = 0.02,\ \lambda_{11}^{XY*} = 0.1,\ \lambda_{11}^{XZ*} = 0.2, \lambda_{12}^{XZ*} = 0.2, \lambda_{11}^{YZ*} = 0.15,\ \lambda_{12}^{YZ*} = 0.15$
(vi) $\lambda_1^{X*} = 0.1,\ \lambda_1^{Y*} = 0.1,\ \lambda_1^{Z*} = 0.1,\ \lambda_2^{Z*} = 0.1,\ \lambda_{11}^{XY*} = 0.2,\ \lambda_{11}^{XZ*} = 0.2,\ \lambda_{12}^{XZ*} = 0.2,\ \lambda_{11}^{YZ*} = 0.2,\ \lambda_{12}^{YZ*} = 0.2$
(vii) $\lambda_1^{X*} = 0.1,\ \lambda_1^{Y*} = 0.1,\ \lambda_1^{Z*} = 0.2,\ \lambda_2^{Z*} = 0.2,\ \lambda_{11}^{XY*} = 0.2,\ \lambda_{11}^{XZ*} = 0.2,\ \lambda_{12}^{XZ*} = 0.1,\ \lambda_{11}^{YZ*} = 0.1,\ \lambda_{12}^{YZ*} = 0.1$
(viii) $\lambda_1^{X*} = 0.2,\ \lambda_1^{Y*} = 0.2,\ \lambda_1^{Z*} = 0.2,\ \lambda_2^{Z*} = 0.2,\ \lambda_{11}^{XY*} = 0.1,\ \lambda_{11}^{XZ*} = 0.1,\ \lambda_{12}^{XZ*} = 0.1,\ \lambda_{11}^{YZ*} = 0.1,\ \lambda_{12}^{YZ*} = 0.1$

We generate N_1 multinomial contingency tables using multinomial random vectors with $\pi = \pi^*$ and arrange the tables as $x^s(\ell)$, $\ell = 1, \ldots, N_1$. For each $\ell = 1, \ldots, N_1$, let $T\{x^s(\ell)\}$ be the value of statistic T at $x^s(\ell)$. Let $\chi_f^2(\alpha)$ be the upper α point of the chi-square distribution with degrees of freedom f, and let N_2 be the number of elements of the set of ℓ that satisfies the condition $T\{x^s(\ell)\} \geq \chi_{N-p-1}^2(\alpha)$. Then the performance of the approximation for the distribution can be evaluated on the basis of the index $I(\alpha) = N_2/N_1 - \alpha$. Tables 7.1, 7.2, 7.3, and 7.4 show the values of $I(\alpha) \times 10^5$ when the true parameters are (i), (ii), (iii), and (iv) and significance levels are α = 0.01 and 0.05 for the log likelihood ratio statistic G^2 and the Bartlett-type transformed statistic G_B^2 in a 2×2×2 contingency table, respectively.

Tables 7.5, 7.6, 7.7, and 7.8 show the values of $I(\alpha) \times 10^5$ when the true parameters are (v), (vi), (vii), and (viii) in a 2×2×3 contingency table, respectively.

In the simulation, the number N_1 of repetitions is 10^6 and the sample sizes are $n = cN$ $(c = 3, 4, 6)$.

From Tables 7.1–7.8, we find that the transformed statistic G_B^2 performs better than the original log likelihood ratio statistic G^2. The results of a comparison indicate that the log likelihood ratio statistic G^2 is improved by the Bartlett-type transformed statistic G_B^2.

Next, we compare the power of the Bartlett-type transformed statistic G_B^2 with that of the log likelihood ratio statistic G^2. Against null model π^*, we consider the alternative model

$$\pi^{**} = \pi^* + \delta, \tag{7.11}$$

Table 7.1 Values of $I(\alpha) \times 10^5$ for G^2 and G_B^2 in the case of a loglinear model in a 2×2×2 contingency table for true parameter (i) with sample size $n = 24, 32, 48$ when significance level α = 0.01 and 0.05

(i)	$n = 24$		$n = 32$		$n = 48$	
α	0.01	0.05	0.01	0.05	0.01	0.05
G^2	1275	3965	1013	3076	631	1700
G_B^2	123	822	179	1038	181	452

Table 7.2 Values of $I(\alpha) \times 10^5$ for G^2 and G_B^2 in the case of a loglinear model in a 2×2×2 contingency table for true parameter (ii) with sample size $n = 24, 32, 48$ when significance level α = 0.01 and 0.05

(ii)	$n = 24$		$n = 32$		$n = 48$	
α	0.01	0.05	0.01	0.05	0.01	0.05
G^2	1110	3865	1085	3637	799	2173
G_B^2	-251	25	76	779	216	615

Table 7.3 Values of $I(\alpha) \times 10^5$ for G^2 and G_B^2 in the case of a loglinear model in a 2×2×2 contingency table for true parameter (iii) with sample size $n = 24, 32, 48$ when significance level α = 0.01 and 0.05

(iii)	$n = 24$		$n = 32$		$n = 48$	
α	0.01	0.05	0.01	0.05	0.01	0.05
G^2	1236	4031	1097	3339	719	2100
G_B^2	-980	412	120	910	160	642

Table 7.4 Values of $I(\alpha) \times 10^5$ for G^2 and G_B^2 in the case of a loglinear model in a 2×2×2 contingency table for true parameter (iv) with sample size $n = 24, 32, 48$ when significance level α = 0.01 and 0.05

(iv)	$n = 24$		$n = 32$		$n = 48$	
α	0.01	0.05	0.01	0.05	0.01	0.05
G^2	1056	3991	1054	3445	777	2022
G_B^2	-274	3	0	694	152	512

Table 7.5 Values of $I(\alpha) \times 10^5$ for G^2 and G_B^2 in the case of a loglinear model in a 2×2×3 contingency table for true parameter (v) with sample size $n = 36, 48, 72$ when significance level α = 0.01 and 0.05

(v)	$n = 36$		$n = 48$		$n = 72$	
α	0.01	0.05	0.01	0.05	0.01	0.05
G^2	1225	4667	1139	4026	729	2269
G_B^2	-15	333	97	953	151	57

Table 7.6 Values of $I(\alpha) \times 10^5$ for G^2 and G_B^2 in the case of a loglinear model in a 2×2×3 contingency table for true parameter (vi) with sample size $n = 36, 48, 72$ when significance level $\alpha = 0.01$ and 0.05

(vi)	$n = 36$		$n = 48$		$n = 72$	
α	0.01	0.05	0.01	0.05	0.01	0.05
G^2	1097	4634	1180	4336	2996	4582
G_B^2	-417	-656	-70	528	784	1428

Table 7.7 Values of $I(\alpha) \times 10^5$ for G^2 and G_B^2 in the case of a loglinear model in a 2×2×3 contingency table for true parameter (vii) with sample size $n = 36, 48, 72$ when significance level $\alpha = 0.01$ and 0.05

(vii)	$n = 36$		$n = 48$		$n = 72$	
α	0.01	0.05	0.01	0.05	0.01	0.05
G^2	1189	4532	1200	4129	909	2804
G_B^2	-313	-367	-31	581	154	704

Table 7.8 Values of $I(\alpha) \times 10^5$ for G^2 and G_B^2 in the case of a loglinear model in a 2×2×3 contingency table for true parameter (viii) with sample size $n = 36, 48, 72$ when significance level $\alpha = 0.01$ and 0.05

(viii)	$n = 36$		$n = 48$		$n = 72$	
α	0.01	0.05	0.01	0.05	0.01	0.05
G^2	1163	4607	1192	4162	893	2823
G_B^2	-366	-447	-26	531	118	796

where $\delta = (\delta_1, \ldots, \delta_N)^\top$, $\sum_{\alpha=1}^{N} \delta_\alpha = 0$ and $0 < \pi_\alpha^* + \delta_\alpha < 1$, $\alpha = 1, \ldots, N$. We calculated the simulated power against the alternative model (7.11) by using the simulated exact critical point of statistics.

Table 7.9 shows the power against the alternative model (7.11) by using the simulated exact critical point when the true parameter is (i), $\delta = (0.16, 0, -0.11, 0.11, 0.09, -0.09, 0, -0.16)^\top$ and significance levels are α =0.01 and 0.05 in a 2×2×2 contingency table. Table 7.10 also shows the power against the alternative model (7.11) when the true parameter is (iv) and $\delta = (0.08, 0, -0.11, 0.11, 0.05, -0.05, 0, -0.08)^\top$. Table 7.11 shows the power against the alternative model (7.11) by using the simulated exact critical point when the true parameter is (v), $\delta = (0.05, 0.06, -0.03, 0.03, 0.05, -0.05, -0.06, -0.05, 0.14, 0.05, -0.05, -0.14)^\top$ and significance levels are α =0.01 and 0.05 in a 2×2×3 contingency table. Table 7.12 also shows the power against the alternative model (7.11) when the true parameter is (vi) and $\delta = (0.06, 0.06, -0.03, 0.03, 0.03, -0.03, -0.06, -0.06, 0.12, 0.01, -0.01, -0.12)^\top$. In the simulation, the number of repetitions is 10^6 and the sample sizes are $n = cN$ $(c = 3, 4, 6)$.

Table 7.9 Simulated power for G^2 and G_B^2 against an alternative model (7.11) for the original loglinear model in a 2×2×2 contingency table for true parameter (i) with sample size n =24, 32, 48 when significance level $\alpha = 0.01$ and 0.05

(i)	$n = 24$		$n = 32$		$n = 48$	
α	0.01	0.05	0.01	0.05	0.01	0.05
G^2	0.04515	0.16474	0.08709	0.25965	0.20338	0.42856
G_B^2	0.04512	0.16474	0.08709	0.25922	0.20360	0.42850

Table 7.10 Simulated power for G^2 and G_B^2 against an alternative model (7.11) for the original loglinear model in a 2×2×2 contingency table for true parameter (iv) with sample size n =24, 32, 48 when significance level $\alpha = 0.01$ and 0.05

(iv)	$n = 24$		$n = 32$		$n = 48$	
α	0.01	0.05	0.01	0.05	0.01	0.05
G^2	0.06703	0.22468	0.12458	0.34492	0.28837	0.55726
G_B^2	0.06482	0.22468	0.12492	0.34461	0.28837	0.55720

Table 7.11 Simulated power for G^2 and G_B^2 against an alternative model (7.11) for the original loglinear model in a 2×2×3 contingency table for true parameter (v) with sample size n =36, 48, 72 when significance level $\alpha = 0.01$ and 0.05

(v)	$n = 36$		$n = 48$		$n = 72$	
α	0.01	0.05	0.01	0.05	0.01	0.05
G^2	0.12152	0.26790	0.21667	0.39769	0.42176	0.62088
G_B^2	0.11984	0.26792	0.21641	0.39749	0.42201	0.62091

Table 7.12 Simulated power for G^2 and G_B^2 against an alternative model (7.11) for the original loglinear model in a 2×2×3 contingency table for true parameter (vi) with sample size n =36, 48, 72 when significance level $\alpha = 0.01$ and 0.05

(vi)	$n = 36$		$n = 48$		$n = 72$	
α	0.01	0.05	0.01	0.05	0.01	0.05
G^2	0.19233	0.36722	0.30059	0.49309	0.50652	0.67578
G_B^2	0.19033	0.36620	0.30082	0.49315	0.50665	0.67585

From Tables 7.9, 7.10, 7.11, and 7.12, we find that the power of G_B^2 is almost the same as the power of G^2. The results for power are as expected since it is known that a transformed statistic by a monotone function that includes the Bartlett adjustment has the same power as that of the original statistic.

7.6 Application to Real Data

We consider the data shown in Table 7.13. The data are from an investigation conducted at the Danish National Institute for Social Research in Copenhagen quoted in Edwards and Kreiner (1983). We extract a 3-way contingency table from original 5-way contingency table.

In the table, 272 skilled workers in different age groups (18–30 years, 31–45 years, and 46–67 years) who lived in houses were cross-classified according to whether they rented or owned their house (Mode: Rent, Own) and whether or not (Response: Yes, No) they had done any work in the preceding 12 months that they would have paid a craftsman to do in earlier times. For the data in Table 7.13, we consider testing whether the 3-factor interaction terms are zero in a loglinear model. The observed value, asymptotic p-value (a.p-value), simulated p-value (s.p-value) that is derived by a more exact distribution by using Monte Carlo simulation and results of the test at a significance level of 0.05 for the log likelihood ratio test statistic G^2 and the Bartlett-type transformed statistic G_B^2 are summarized in Table 7.14.

From Table 7.14, the observed value of the log likelihood ratio statistic G^2 is 6.109 and the value of the Bartlett-type transformed statistic G_B^2 is 5.739. Since $\chi_2^2(0.05)$ = 5.991, the hypothesis that the loglinear model is true cannot be rejected at the 5% level using transformed statistic G_B^2.

Now, the asymptotic p-value of G^2 is 0.047 and the simulated p-value of G^2 is 0.060. On the other hand, the asymptotic p-value of G_B^2 is 0.057 and the simulated p-value of G_B^2 is 0.060. Therefore, for the statistic G^2, the result of the test based on an asymptotic distribution is opposite to the result of the test based on a more exact distribution at the 5% level. In contrast, for the statistic G_B^2, the result of the test based on an asymptotic distribution agrees with the result of the test based on a more exact distribution at the 5% level.

Table 7.13 A 3-way table extracted from Edwards and Kreiner (1983)

		Age		
Mode	Response	18–30	31–45	46–67
Rent	Yes	34	10	2
	No	28	4	6
Own	Yes	56	56	35
	No	12	21	8

Table 7.14 Results for $G^2$and G_B^2 in the data in Table 7.13

	G^2	G_B^2
Observed values of test statistics	6.109	5.739
Asymptotic critical points for the 5% level	5.991	5.991
a.p-values	0.047	0.057
Results of asymptotic test	Reject	Accept
s.p-values	0.060	0.060
Results of the test based on an exact distribution	Accept	Accept

References

Agresti A (2013) Categorical data analysis, 3rd edn. Wiley, New York

Bartlett MS (1935) Contingency table interactions. J R Statist Soc Suppl 2:248–252

Edwards D, Kreiner S (1983) The analysis of contingency tables by graphical models. Biometrika 70:553–565

Kobe T, Taneichi N, Sekiya Y (2015) Improved transformed statistics for the test of one factor independence from the other two in an $r \times s \times t$ contingency table. J Jpn Stat Soc 45(1):77–98

Taneichi N, Sekiya Y, Toyama J (2019) Transformed statistics for tests of conditional independence in $J \times K \times L$ contingency tables. J Multivariate Anal 171:193–208

Taneichi N, Sekiya Y, Toyama J (2021) Improvement of the test of independence among groups of factors in a multi-way contingency table. Jpn J Stat Data Sci 4:181–213

Chapter 8
The Selection of Statistics Based on a Second-Order Correction Term When Data Are Not So Large

Keywords Complete independence · Contingency table · Multinomial goodness-of-fit test · ϕ-divergence statistics · Power divergence statistics · Second-order correction term · Sparse data · Test of complete independence

8.1 Previous Study on the Accuracy of Chi-Square Approximations of Test Statistics

For a multinomial goodness-of-fit test, Larntz (1978) and Lawal (1984) compared the accuracies of the limiting chi-square approximation of distributions of Pearson's X^2 statistic, the log likelihood ratio statistic and the Freeman-Tukey statistic. For a 2×2 contingency table, Upton (1982) also compared the accuracies of the chi-square limiting approximation of the above statistics. They concluded that Pearson's X^2 statistic is more accurate than the log likelihood ratio statistic and the Freeman-Tukey statistic. Furthermore, for a multinomial goodness-of-fit test, Cressie and Read (1984) and Read and Cressie (1988) compared the accuracies of the limiting chi-square approximation of the distribution of the power divergence family of statistics R^a. Their results indicated that $R^{2/3}$ and R^1 (Pearson's X^2 statistic) are very accurate.

This chapter is based on the following article: Japanese Journal of Statistics and Data Science, 7, Taneichi N. and Sekiya Y., Selection of statistics for a multinomial goodness-of-fit test and a test of independence for a multi-way contingency table when data are sparse, 151–182, Springer Nature (2024).

N. Taneichi and Y. Sekiya, *Improving Tests for Discrete Small Sample Data*, JSS Research Series in Statistics,
https://doi.org/10.1007/978-981-95-5301-3_8

8.2 Sparse Data

The assumption that the number k of categories increases to infinity as the number n of sample sizes increases to infinity in a multinomial distribution and the assumption that the number of cells increases to infinity as the number of sample sizes increases to infinity in contingency tables are called the sparseness assumption of a multinomial distribution and the sparseness assumption of contingency tables, respectively. Asymptotic normality theories and the numerical performance of multinomial models and contingency tables under the sparseness assumption have been investigated by various authors (Holst 1972; Morris 1975; Dale 1986; Koehler 1986; etc.). For example, Holst (1972) proved that R^0 and R^1 have an asymptotic normal distribution when n and k tend to infinity so that $n/k \to \beta$ $(0 < \beta < \infty)$. Notably, the asymptotic distribution of statistics for disparity in product multinomial models based on ϕ -divergence under the sparseness assumption was investigated in Morales et al. (2003).

In a practical problem, the sparseness assumption symbolically represents cases in which there are few samples per category or per cell. We call such data sparse data. In this chapter, for sparse data, that is, data for which there are only a few samples per category or per cell, we do not consider the asymptotic normality of test statistics under the sparseness assumption, but we look for the test statistics for which the distribution is well approximated by the chi-square limiting distribution when $n \to \infty$. For that reason, by using the second-order correction term, which will be shown in Sect. 8.3, as a basis, we consider an index to choose the statistics for a goodness-of-fit test for a multinomial distribution, a test of complete independence, and a test of independence among groups of factors in a multi-way contingency table.

8.3 Second-Order Correction Term

For the goodness-of-fit test of a multinomial distribution with number k of categories, the null distribution of test statistic C_ϕ is asymptotically distributed as a chi-square distribution as $n \to \infty$ with k fixed. Usually, when the sample size is large to some degree, the hypothesis is tested by using the chi-square distribution.

We investigate the characteristics of function ϕ that gives the desirable test statistic C_ϕ for data for which the number of samples per category, that is, n/k, is comparatively small. As a first step, we consider the case in which n is large with k fixed. We consider the statistic T_ϕ for which we can evaluate the sth moments about the origin of the statistic as

$$E\{(T_\phi)^s|H_0\} = E((F_\lambda)^s) + \frac{m_\phi^T(s)}{n} + o(n^{-1}), \quad s = 1, 2, \ldots, \tag{8.1}$$

where F_λ is a random variable distributed according to a chi-square distribution with degrees of freedom λ and $m_\phi^T(s)$ is a second-order correction term.

From the definition of the second-order correction term $m_\phi^T(s)$ by (8.1), $m_\phi^T(s)$ is considered as an evaluation of the difference between the sth, $s = 1, 2, \dots$ moment of the distribution of statistic T_ϕ when n is large and that of the chi-square distribution, that is, the difference between the distribution of statistic T_ϕ when n is large and the chi-square distribution. If the absolute value of the second-order correction term is close to zero, the distribution of the statistic is considered to be close to the chi-square distribution with degrees of freedom λ. Therefore, we consider the second-order correction term $m_\phi^T(s)$ as an index for investigating whether the null distribution of statistic T_ϕ is close to the chi-square limiting distribution with degrees of freedom λ. When T_ϕ in (8.1) is C_ϕ, the second-order correction term $m_\phi^C(s)$ is given by (A.43) in Appendix A.9. It is a polynomial of degree $s-1$ of λ. Coefficients of the polynomial include $\phi'''(1)$ and $\phi^{(4)}(1)$. A test statistic for which the null distribution is close to the chi-square distribution is defined by the function ϕ that satisfies $m_\phi^C(s) = 0$. Such a function ϕ depends on k and the null hypothesis, that is, $q = (q_1, \dots, q_k)^\top$ given by (2.1). We premise to test the hypothesis by using the chi-square limiting distribution for data for which n/k are comparatively small (sparse data). Then, as a second step, in order to obtain the test statistic for which the distribution is approximated well to the chi-square distribution for sparse data, we look for the function ϕ for which $m_\phi^C(s)$ approaches 0 when $k \to \infty$. The test statistic constructed by the function ϕ that is given above is considered to be effective for analyzing sparse data.

8.4 Theorems for Multinomial Goodness-of-Fit Test Statistics

In this section, we derive theorems for the multinomial goodness-of-fit test. We consider the moment of statistics C_ϕ given by (2.2) under the null hypothesis H_0 given by (2.1). With regard to the second-order correction term $m_\phi^C(s)$, we obtain the following theorem.

Theorem 8.1 *Let ϕ be four times continuously differentiable. Assuming that the distribution of ϕ-divergence statistic C_ϕ is continuous and that $q_j = O(k^{-1})$, $j = 1, \dots, k$, then for ϕ-divergence statistic C_ϕ having ϕ that satisfies $m_\phi^C(s) = 0$, $s = 1, 2, \dots,$ $4\phi'''(1) + 3\phi^{(4)}(1)$ tends to zero as k tends to infinity under H_0.*

The proof of Theorem 8.1 is shown in Appendix A.9. The special null hypothesis that is denoted by $q = (1/k, \dots, 1/k)^\top$ in hypothesis (2.1) is called the symmetric null hypothesis. If we put

$$S = \sum_{j=1}^{k} q_j^{-1}, \tag{8.2}$$

it is clear that $S = k^2$ holds if and only if the hypothesis is symmetric. The results of Theorem 8.1 are shown in the cases of $s = 1, 2$, and 3 under the symmetric null hypothesis in Pardo (1988, p. 183). Theorem 8.1 extends the results of Pardo (1988, p. 183) to a certain kind of general hypothesis that includes the symmetric hypothesis and also extends the results of Pardo (1988, p. 183) for arbitrary natural number s.

We note that the theorem holds in all sth, $s = 1, 2, \ldots$ moments of statistics, and we can therefore say that the property of the theorem holds not only in the moments of statistics but also in the distribution of statistics.

When the statistic is the power divergence statistic R^a, we substitute ϕ^a in (2.3) for ϕ. Since $\phi^{a'''}(1) = a - 1$ and $\phi^{a(4)}(1) = (a-1)(a-2)$, by applying the results to Theorem 8.1, we obtain the following corollary.

Corollary 8.1 *Assuming that the distribution of the power divergence statistic R^a is continuous and that $q_j = O(k^{-1}),\ j = 1, \ldots, k$, then for the power divergence statistic R^a, two roots of the equation $m^R_{\phi^a}(s) = 0,\ s = 1, 2, \ldots$ tend to $a = 1$ and $a = 2/3$ as k tends to infinity under H_0.*

The result of Corollary 8.1 generalizes the results of Read and Cressie (1988) and Pardo (1988, p. 183). When the statistic is Rukhin's statistic Q^a, we substitute ϕ^a_Q in (2.4) for ϕ. Since $\phi^{a\,'''}_Q(1) = 3(a-1)$ and $\phi^{a(4)}_Q(1) = 12(a-1)^2$, by applying the results to Theorem 8.1, we obtain the following corollary.

Corollary 8.2 *Assuming that the distribution of Rukhin's statistic Q^a is continuous and that $q_j = O(k^{-1}),\ j = 1, \ldots, k$, then for Rukhin's statistic Q^a, two roots of the equation $m^Q_{\phi^a_Q}(s) = 0,\ s = 1, 2, \ldots$ tend to $a = 1$ and $a = 2/3$ as k tends to infinity under H_0.*

We obtain the following theorem with regard to the evaluation of $m^C_\phi(s)$.

Theorem 8.2 *Let ϕ be four times continuously differentiable. Assuming that the distribution of ϕ-divergence statistic C_ϕ is continuous and that $q_j = O(k^{-1}),\ j = 1, \ldots, k$, then for ϕ that satisfies*

$$4\phi'''(1) + 3\phi^{(4)}(1) = 0, \tag{8.3}$$

$m^C_\phi(s)$ is evaluated as

$$\begin{aligned} m^C_\phi(s) = \frac{s}{6}\Big[&(s-1)\left\{5(\phi'''(1)+1)^2 - 2\right\}\left(\frac{S}{k^2}\right) \\ &-\Big\{3(s-1)(\phi'''(1)+1)^2 + 2\phi'''(1)\Big\}\Big]k^s \\ &+O(k^{s-1}),\quad s = 1, 2, \ldots \end{aligned}$$

under H_0, where S is defined in (8.2).

The proof of Theorem 8.2 is shown in Appendix A.9. Using this theorem, we obtain the following corollary about the statistics R^1 and $R^{2/3}$.

Corollary 8.3 *Assuming that the distribution of the power divergence statistic R^a is continuous and that $q_j = O(k^{-1}),\ \ j = 1, \dots, k$, then second-order correction terms of the sth moments of the statistics R^1 and $R^{2/3}$ are evaluated as*

$$m^R_{\phi^1}(s) = A_s k^s + O(k^{s-1}), \quad s = 1, 2, \dots \tag{8.4}$$

and

$$m^R_{\phi^{2/3}}(s) = B_s k^s + O(k^{s-1}), \quad s = 1, 2, \dots, \tag{8.5}$$

where

$$A_s = \frac{1}{2}s(s-1)\left(\frac{S}{k^2} - 1\right), \quad s = 1, 2, \dots$$

and

$$B_s = \frac{s}{27}\left\{(s-1)\frac{S}{k^2} - 3(2s-3)\right\}, \quad s = 1, 2, \dots$$

under H_0, respectively. Therefore, $|A_1| < |B_1|$. Furthermore, $|A_s| < |B_s|\ (s \geq 2)$ when $1 \leq S/k^2 < C(s)$ and $|A_s| > |B_s|\ (s \geq 2)$ when $C(s) < S/k^2$, where

$$C(s) = \frac{3(13s - 15)}{29(s-1)}.$$

The proof of Corollary 8.3 is shown in Appendix A.10. Expressions (8.4) and (8.5) in Corollary 8.3 are represented as $m^R_{\phi^1}(s)/k^{s+1} = A_s/k + O(k^{-2})$ and $m^R_{\phi^{2/3}}(s)/k^{s+1} = B_s/k + O(k^{-2})$. Therefore, we can compare the speed of convergence of $m^R_{\phi^1}(s)/k^{s+1}$ to 0 as $k \to \infty$ with that of $m^R_{\phi^{2/3}}(s)/k^{s+1}$ by comparing the sizes of $|A_s|$ and $|B_s|$. So, from Corollary 8.3, we find that $m^R_{\phi^1}(s)/k^{s+1}$ $(s \geq 1)$ converges to zero faster than $m^R_{\phi^{2/3}}(s)/k^{s+1}$ does as k increases when $1 \leq S/k^2 < C(s)$. Since $S/k^2 = 1$ represents a case of a symmetric hypothesis, this result also holds under a symmetric null hypothesis. On the other hand, we find that $m^R_{\phi^{2/3}}(s)/k^{s+1}$ $(s \geq 2)$ converges to zero faster than $m^R_{\phi_1}(s)/k^{s+1}$ does as k increases when $C(s) < S/k^2$.

From the above discussion, for choosing a test statistic for which the distribution converges to the chi-square limiting distribution under H_0 more quickly among the statistics based on ϕ-divergence, we recommend test statistic C_ϕ for testing H_0 where ϕ satisfies (8.3). Among power divergence statistics R^a, we recommend R^1 (Pearson's X^2 statistic) or $R^{2/3}$ for testing H_0. Among Rukhin's statistics Q^a, we recommend Q^1 (Pearson's X^2 statistic) and $Q^{2/3}$ for testing H_0.

8.5 Theorems for Test of Complete Independence in a Contingency Table

In this section, we extend the theorems for a goodness-of-fit test for a simple null hypothesis to theorems for a test of complete independence in multi-way contingency tables. We consider M-way $J_1 \times J_2 \times \cdots \times J_M$ contingency tables as a full multinomial model in which sample size n is fixed as stated in Sect. 4.6. We also consider test statistic $C_{\phi}^{(1)[M]}$ defined by (4.13) for testing hypothesis H_0^M given by (4.11). Hereafter, we write M_{ϕ} instead of $C_{\phi}^{(1)[M]}$ for simplicity.

Then, if ϕ is four times continuously differentiable, the test statistic $M_{\phi}(U)$ can be expanded as follows:

$$M_{\phi}(U) = M_{\phi}^*(U) + o_p(n^{-3/2}),$$

where

$$M_{\phi}^*(U) = \tau_0(U) + n^{-1/2}\tau_1^{\phi}(U) + n^{-1}\tau_2^{\phi}(U), \tag{8.6}$$

$\tau_0(U)$, $\tau_1^{\phi}(U)$, and $\tau_2^{\phi}(U)$ are given by substituting U for u in (A.25) of Appendix A.4, respectively.

When we consider the test of complete independence in multi-way contingency tables, there are many cases in which moments of the statistic M_{ϕ} do not exist. However, similar to the reason for example 14.6-1 of Bishop et al. (1975, p. 488), since the probability that $(M_{\phi})^s = \infty$ goes to zero very quickly, for a large number n of samples, the distribution of M_{ϕ} is thought to be very close to the distribution of M_{ϕ}^*. As for the test statistics for complete independence in M-way contingency tables, we consider the moments of M_{ϕ}^* instead of M_{ϕ}.

Since statistics M_{ϕ}^* are asymptotically distributed as a chi-square distribution with degrees of freedom $\mu = f^{[M]}$, where $f^{[M]}$ is given by (4.16) under H_0^M as well as M_{ϕ}, we can evaluate the sth moment about the origin of M_{ϕ}^* as

$$E\{(M_{\phi}^*)^s|H_0^M\} = E((F_{\mu})^s) + \frac{m_{\phi}^M(s)}{n} + o(n^{-1}), \quad s = 1, 2, \ldots.$$

Therefore, we can define the second-order correction term $m_{\phi}^M(s)$. With regard to the second-order correction term $m_{\phi}^M(s)$, we obtain the following theorem.

Theorem 8.3 *Let ϕ be four times continuously differentiable, and put $K = \prod_{m=1}^{M} J_m$. Assume that the* distribution of the expanded expression *M_{ϕ}^* given by (8.6) is continuous, that $p_{j_1 \ldots j_M} = O(K^{-1})$, $j_m = 1, \ldots, J_m$; $m = 1, \ldots, M$, and that $O(J_1) = \cdots = O(J_M)$. Then, for ϕ that satisfies $m_{\phi}^M(s) = 0$, $s = 1, 2, \ldots$, where $m_{\phi}^M(s)$ is the second-order correction term of expanded expression M_{ϕ}^*,*

$4\phi'''(1) + 3\phi^{(4)}(1)$ *tends to zero as* K *tends to infinity under* H_0^M.

The proof of Theorem 8.3 is shown in Appendix A.11. Let R_M^{a*} be the expanded expression given by (8.6) for the statistic $R_M^a \equiv M_{\phi^a}$ based on power divergence, that is, $M_{\phi^a}^*(U)$. For R_M^{a*}, we obtain the following corollary.

Corollary 8.4 *Assume that the distribution of the expanded expression* R_M^{a*} *is continuous, that* $p_{j_1 \ldots j_M} = O(K^{-1})$, $j_m = 1, \ldots, J_m$; $m = 1, \ldots, M$, *and that* $O(J_1) = \cdots = O(J_M)$. *Then, two roots of the equation* $m_{\phi^a}^M(s) = 0,\ s = 1, 2, \ldots$ *tend to* $a = 1$ *and* $a = 2/3$ *as* K *tends to infinity under* H_0^M.

Let Q_M^{a*} be the expanded expression given by (8.6) for Rukhin's statistic $Q_M^a \equiv M_{\phi_Q^a}$, that is, $M_{\phi_Q^a}^*(U)$. For Q_M^{a*}, we obtain the following corollary.

Corollary 8.5 *Assume that the distribution of the expanded expression* Q_M^{a*} *is continuous, that* $p_{j_1 \ldots j_M} = O(K^{-1})$, $j_m = 1, \ldots, J_m$; $m = 1, \ldots, M$, *and that* $O(J_1) = \cdots = O(J_M)$. *Then, two roots of the equation* $m_{\phi_Q^a}^M(s) = 0,\ s = 1, 2, \ldots$ *tend to* $a = 1$ *and* $a = 2/3$ *as* K *tends to infinity under* H_0^M.

If we put

$$S_m = \sum_{j_m=1}^{J_m} p_{\cdot(m, j_m)}^{-1}, \quad m = 1, \ldots, M, \tag{8.7}$$

we obtain the following theorem with regard to the evaluation of $m_\phi^M(s)$.

Theorem 8.4 *Let* ϕ *be four times continuously differentiable. Assume that the distribution of the expanded expression* M_ϕ^* *given by (8.6) is continuous, that* $p_{j_1 \ldots j_M} = O(K^{-1})$, $j_m = 1, \ldots, J_m$; $m = 1, \ldots, M$, *and that* $O(J_1) = \cdots = O(J_M)$. *Then, for* ϕ *that satisfies (8.3),* $m_\phi^M(s)$ *is evaluated as*

$$m_{\phi}^{M}(s) = -\frac{s}{3}\phi'''(1)K^{s}\sum_{m=1}^{M}\frac{S_m}{J_m} + O\left(K^{s}\right) + O\left(K^{s-1}\left(\sum_{m=1}^{M}J_m\right)^{2}\right), \quad s = 1, 2, \ldots$$

under H_0^M.

The proof of Theorem 8.4 is shown in Appendix A.11.

From Theorem 8.4, if we would like to choose a test statistic for which the distribution converges to the chi-square limiting distribution under H_0^M more quickly among the statistics based on ϕ-divergence, we recommend test statistic M_ϕ for testing H_0^M where ϕ satisfies (8.3) and $\phi'''(1) = 0$, that is, $\phi'''(1) = \phi^{(4)}(1) = 0$. Therefore, among power divergence statistics R_M^a, we recommend R_M^1 (Pearson's X^2 statistic) for testing H_0^M. Among Rukhin's statistics Q_M^a, we recommend Q_M^1 (Pearson's X^2 statistic) for testing H_0^M.

On the other hand, a test for two factors independence from another in a 3-way $J_1 \times J_2 \times J_3$ contingency table is reduced to a test of independence in a 2-way $J_1 J_2 \times J_3$ contingency table. In general, a test for independence among groups of factors in an M-way contingency table is reduced to a test of complete independence in an N-way contingency table, where N is less than M. Therefore, in the case of a test of independence among groups of factors in a multi-way contingency table, we obtain the same recommendation as that in the case of the test of complete independence.

References

Bishop YMM, Fienberg SE, Holland PW (1975) Discrete multivariate analysis: theory and practice. MIT Press, Cambridge

Cressie N, Read TRC (1984) Multinomial goodness-of-fit tests. J Roy Statist Soc Ser B 46:440–464

Dale JR (1986) Asymptotic normality of goodness-of-fit statistics for sparse product multinomials. J R Stat Soc Ser B 48:48–59

Holst L (1972) Asymptotic normality and efficiency for certain goodness-of-fit tests. Biometrika 59:137–145

Koehler KJ (1986) Goodness-of-fit tests for log-linear models in sparse contingency tables. J Am Stat Assoc 81:483–493

Larntz K (1978) Small-sample comparisons of exact levels for chi-squared goodness-of-fit statistics. J Am Stat Assoc 73:253–263

Lawal HB (1984) Comparisons of the X^2, Y^2, Freeman-Tukey and Williams's improved G^2 test statistics in small samples of one-way multinomials. Biometrika 71:415–418

Morales D, Pardo L, Vajda I (2003) Asymptotic laws for disparity statistics in product multinomial models. J Multivariate Anal 85:335–360

Morris C (1975) Central limit theorems for multinomial sums. Ann Stat 3:165–188

Pardo L (2006) Statistical inference based on divergence measures. Chapman & Hall/CRC

Read TRC, Cressie NAC (1988) Goodness-of-fit statistics for discrete multivariate data. Springer, New York

Upton GJG (1982) A comparison of alternative tests for the 2 × 2 comparative trial. J R Stat Soc Ser A 145:86–105

Chapter 9
Constructing a New Statistic NT for Testing Complete Independence in Contingency Tables

Keywords M-way contingency table · Pearson's X^2 statistic · ϕ-divergence statistics · power divergence statistics · Test of complete independence

9.1 A New Statistic in Addition to Pearson's X^2 Statistic Satisfying Conditions for Suitability

We consider the conditions (9.1) of ϕ as follows.

$$\phi(1) = \phi'(1) = \phi'''(1) = \phi^{(4)}(1) = 0, \ \phi''(1) = 1, \ \text{and } \phi(t) \text{ is convex on } t > 0. \tag{9.1}$$

Then, from Theorems 8.3 and 8.4, test statistic M_ϕ defined by ϕ satisfying the conditions of (9.1) is considered as a test statistic for which the distribution quickly converges to a chi-square distribution. For the family of power divergence statistics, only R_M^1 (Pearson's X^2 statistic) is defined by the ϕ satisfying conditions (9.1). For the family proposed by Rukhin (1994), only Q_M^1 (Pearson's X^2 statistic) is defined by the ϕ satisfying conditions (9.1) as well.

In this section, we seek a new ϕ satisfying conditions (9.1). By using ϕ, we obtain a new test statistic for complete independence for which the distribution quickly converges to a chi-square distribution. In order to obtain the statistic, we consider as follows.

This chapter is based on the following article: Journal of the Japan Statistical Society Ser. J 55(1), Taneichi N. and Sekiya Y., A new test statistic based on ϕ-divergence for testing independence of contingency tables that minimizes the second-order correction term (in Japanese), 1–24, (2025). The copyright of this article belongs to the Japan Statistical Society. Permission to reuse the material has been obtained from the Japan Statistical Society.

N. Taneichi and Y. Sekiya, *Improving Tests for Discrete Small Sample Data*,
JSS Research Series in Statistics,
https://doi.org/10.1007/978-981-95-5301-3_9

When we consider the functions ϕ^a and ϕ_Q^a given by (2.3) and (2.4), respectively, both $\phi^{2/3}$ and $\phi_Q^{2/3}$ satisfy the condition $4\phi'''(1) + 3\phi^{(4)}(1) = 0$. Then the linear combination

$$\psi(t) = \alpha\phi^{2/3}(t) + \beta\phi_Q^{2/3}(t) \ (t \geq 0) \ (\alpha, \beta \text{ being arbitrary})$$

also satisfies $4\psi'''(1) + 3\psi^{(4)}(1) = 0$. On the other hand, for any a, $\phi^{a''}(1) = \phi_Q^{a''}(1) = 1$, $\phi^{a'''}(1) = a - 1$, and $\phi_Q^{a'''}(1) = 3(a-1)$. Therefore, if $\psi''(1) = 1$ is satisfied, then $\alpha + \beta = 1$ is required. Also, if $\psi'''(1) = 0$ is satisfied, then $\alpha + 3\beta = 0$ is required. This means that $\psi''(1) = 1$ and $\psi'''(1) = 0$ when $\alpha = 3/2$ and $\beta = -1/2$. If we put

$$\phi_{NT}(t) \equiv \frac{3}{2}\phi^{2/3}(t) - \frac{1}{2}\phi_Q^{2/3}(t) = \frac{27}{20}t^{5/3} - 3t - \frac{27}{4}(t+2)^{-1} + \frac{39}{10}, \qquad (9.2)$$

then $\phi''_{NT}(1) = 1$, and $\phi'''_{NT}(1) = \phi^{(4)}_{NT}(1) = 0$ are satisfied. Furthermore, $\phi_{NT}(t)$ is convex on $t > 0$ as shown in Appendix A.12. Since $\phi_{NT}(t)$ satisfies the conditions (9.1), we can construct a test statistic for which the distribution quickly converges to a chi-square distribution as well as Pearson's X^2 statistic by using $\phi_{NT}(t)$. We call this statistic the NT statistic (New Test Statistic).

For the hypothesis of complete independence of an M-way $J_1 \times J_2 \times \cdots \times J_M$ contingency table, by substituting ϕ_{NT} for ϕ in (4.13), the constructed test statistic (NT statistic) is

$$C_{\phi_{NT}} = \frac{27n}{10}\sum_{j_1=1}^{J_1}\cdots\sum_{j_M=1}^{J_M}\left\{\hat{p}_{j_1\ldots j_M}\left(\frac{\hat{p}_{j_1\ldots j_M}}{\hat{Q}(j_1\ldots j_M)}\right)^{2/3} - \frac{5\hat{Q}(j_1\ldots j_M)^2}{\hat{p}_{j_1\ldots j_M} + 2\hat{Q}(j_1\ldots j_M)}\right\} + \frac{9}{5}n,$$

where $\hat{Q}(j_1\ldots j_M)$ is given by (4.14), and $\hat{p}_{j_1\ldots j_M}$ is given by (4.15).

Especially, for a 2-way $J \times K$ contingency table X_{jk}, $j = 1, \ldots, J$, $k = 1, \ldots, K$, the test statistic $C_{\phi_{NT}}$ for null hypothesis $H_0^{(I)} : p_{jk} = p_{j\cdot}p_{\cdot k}$, $j = 1, \ldots, J, k = 1, \ldots, K$ is

$$C_{\phi_{NT}} = \frac{27}{10}\sum_{j=1}^{J}\sum_{k=1}^{K}X_{jk}\left(\frac{nX_{jk}}{X_{j\cdot}X_{\cdot k}}\right)^{2/3} - \frac{27}{2}\sum_{j=1}^{J}\sum_{k=1}^{K}\frac{\left(\frac{X_{j\cdot}X_{\cdot k}}{n}\right)^2}{X_{jk} + 2\left(\frac{X_{j\cdot}X_{\cdot k}}{n}\right)} + \frac{9}{5}n.$$

Furthermore, for a 3-way $J \times K \times L$ contingency table given by Table 4 in Chap. 4, the test statistic $C_{\phi_{NT}}$ for the complete independence null hypothesis $H_0^{(1)} : p_{jk\ell} = p_{j\cdot\cdot}p_{\cdot k\cdot}p_{\cdot\cdot\ell}$, $j = 1, \cdots, J, k = 1, \cdots, K, \ell = 1, \cdots, L$ is

$$C_{\phi_{NT}} = \frac{27}{10} \sum_{j=1}^{J} \sum_{k=1}^{K} \sum_{\ell=1}^{L} X_{jk\ell} \left(\frac{n^2 X_{jk\ell}}{X_{j\cdot\cdot} X_{\cdot k \cdot} X_{\cdot\cdot\ell}} \right)^{2/3}$$

$$- \frac{27}{2} \sum_{j=1}^{J} \sum_{k=1}^{K} \sum_{\ell=1}^{L} \frac{\left(\frac{X_{j\cdot\cdot} X_{\cdot k \cdot} X_{\cdot\cdot\ell}}{n^2} \right)^2}{X_{jk\ell} + 2 \left(\frac{X_{j\cdot\cdot} X_{\cdot k \cdot} X_{\cdot\cdot\ell}}{n^2} \right)} + \frac{9}{5} n.$$

In Sect. 9.2, the performance of the approximation by a chi-square distribution of the NT statistic is compared with that of the other statistics including Pearson's X^2 statistic.

9.2 Numerical Comparison of the NT Statistic and the Other Statistics

In this section, we investigate the performance of an approximation by a chi-square distribution of the distribution of the NT statistic when the sample size is not so large. For a test of independence of a 2-way $J \times K$ contingency table, complete independence of a 3-way $J \times K \times L$ contingency table and a 4-way $J \times K \times L \times M$ contingency table, we compare the approximation by a chi-square distribution of power divergence statistics $R_M^a \equiv M_{\phi^a}$ including Pearson's X^2 statistic (R_M^1) with that of the NT statistic.

We evaluate the performance using the following procedure.
(1) We generate N_1 multinomial contingency tables using multinomial random vectors under H_0 and arrange the tables as $x^S(h),\ h = 1, \ldots, N_1$.
(2) Let $T(x^S(h)), h = 1, \ldots, N_1$ be the value of a statistic T (R_M^a or $C_{\phi_{NT}}$) at $x^S(h)$.
(3) Let $\chi_\nu^2(\alpha)$ be the α upper point of the chi-square distribution degrees of freedom ν. For test statistics of complete independence R_M^a and $C_{\phi_{NT}}$, let $\nu = JK - (J + K) + 1$ for a 2-way contingency table, let $\nu = JKL - (J + K + L) + 2$ for a 3-way contingency table, and let $\nu = JKLM - (J + K + L + M) + 3$ for a 4-way contingency table.
(4) Let N_2 be the number of elements of the set of h that satisfies the condition $T(x^S(h)) \geq \chi_\nu^2(\alpha)$.

Then, the performance of the approximation for the distribution can be evaluated on the basis of the index $|I(\alpha)|$, where $I(\alpha) = N_2/N_1 - \alpha$ and $|a|$ is the absolute value of a.

First, we consider 2-way contingency tables. For a 2-way $J \times K$ contingency table, the following sets of marginal probabilities are considered.

Case (I): All marginal probabilities are equal, that is,

$$p_{j\cdot} = J^{-1}, \quad j = 1, \ldots, J, \quad p_{\cdot k} = K^{-1}, \quad k = 1, \ldots, K.$$

Case (II): Marginal probabilities $p_{j\cdot}$ and $p_{\cdot k}$ increase linearly corresponding to j, k, that is,

$$p_{j\cdot} = \frac{1 + J + 2j}{2J(J+1)}, \quad j = 1, \ldots, J, \quad p_{\cdot k} = \frac{1 + K + 2k}{2K(K+1)}, \quad k = 1, \ldots, K.$$

Case (III): Marginal probabilities $p_{j\cdot}$ and $p_{\cdot k}$ are similar to those used for the investigation of a sparse contingency table by Koehler (1986)

$$p_{j\cdot} = J^{-1}\left(0.1 + 0.9\sum_{m=j}^{J} m^{-1}\right), \quad j = 1, \ldots, J,$$

$$p_{\cdot k} = K^{-1}\left(0.1 + 0.9\sum_{m=k}^{K} m^{-1}\right), \quad k = 1, \ldots, K.$$

We define the sparseness index SI of a contingency table as

$$SI = \frac{n}{(\text{Total number of cells})},$$

where n is the sample size.

The values of $I(\alpha) \times 10^5$ for statistics R_M^a $(a = 0, 2/3, 1)$ and $C_{\phi_{NT}}$, provided that $N_1 = 10^6$ when the level of significance $\alpha = 0.01, 0.05$ and index $SI = 2.0 \sim 4.0$ for 4×4, 5×5, and 6×6 contingency tables are shown in Tables 9.1, 9.2, and 9.3, respectively.

In addition to Case (I)–Case (III), we consider the real data shown in Table 9.4. Table 9.4 shows the relationship between the number of nonprofit health organizations among the top ten employers in a city and region of the country for 72 of the largest cities in the United States. The data were obtained by Abzug et al. (2000) and are quoted in Simonoff (2003, pp. 240–241). We consider a model by using the estimated marginal probabilities $p_{j\cdot}$, $j = 1, \ldots, 4$ and $p_{\cdot k}$, $k = 1, \ldots, 5$ from the real data given by Table 9.4 as true values. Table 9.5 shows the results of the simulation similar to Case (I)–Case (III) for the model.

From Tables 9.1, 9.2, and 9.3, by seeing the absolute value of the approximation error $|I(\alpha)|$, we find that the distributions of the NT statistic and Pearson's X^2 statistic are close to a chi-square limiting distribution, and a chi-square approximation is considered to perform well even when the sample size is not so large. From Table 9.5, for the model based on real data, the NT statistic and Pearson's X^2 statistic also perform well.

Next, we consider a comparison between Pearson's X^2 statistic and the NT statistic. In Tables 9.1, 9.2, and 9.3, when we compare the values of $|I(\alpha)|$ for Pearson's

Table 9.1 Values of $I(\alpha) \times 10^5$ for 4×4 tables in Cases (I)–(III) for statistics R_M^a ($a =$ 0.0, 2/3, 1.0) and $C_{\phi NT}$ when $\alpha = 0.01$ and 0.05 and sparseness index SI = 2.0, 2.5, 3.0, and 4.0

4 × 4		$SI = 2.0$		$SI = 2.5$		$SI = 3.0$		$SI = 4.0$	
Case	$a \backslash \alpha$	0.01	0.05	0.01	0.05	0.01	0.05	0.01	0.05
(I)	0.0	1500	6435	1555	5866	1407	5080	1192	3542
	2/3	–274	–388	–218	–142	–131	–15	–56	161
	1.0	–239	–547	–186	–425	–159	–297	–102	–159
	NT	–179	–253	–136	–132	–103	–86	–47	19
(II)	0.0	1169	5499	1277	5225	1360	4812	999	3731
	2/3	–421	–724	–316	–487	–190	–166	–157	–108
	1.0	–289	–675	–226	–612	–168	–365	–183	–272
	NT	–267	–467	–197	–374	–126	–155	–143	–96
(III)	0.0	–293	12	–44	985	159	1560	306	2075
	2/3	–560	–1990	–448	–1565	–397	–1216	–268	–915
	1.0	737	320	425	–25	188	–35	134	–164
	NT	–70	–773	–59	–444	–4	–290	45	–218

Table 9.2 Values of $I(\alpha) \times 10^5$ for 5×5 tables in Cases (I)–(III) for statistics R_M^a ($a =$ 0.0, 2/3, 1.0) and $C_{\phi NT}$ when $\alpha = 0.01$ and 0.05 and sparseness index SI = 2.0, 2.48, 3.0, and 4.0

5 × 5		$SI = 2.0$		$SI = 2.48$		$SI = 3.0$		$SI = 4.0$	
Case	$a \backslash \alpha$	0.01	0.05	0.01	0.05	0.01	0.05	0.01	0.05
(I)	0.0	2039	8400	2114	7527	1922	6295	1324	4428
	2/3	–357	–547	–245	–199	–204	91	–96	58
	1.0	–282	–645	–206	–422	–180	–208	–122	–190
	NT	–224	–295	–140	–87	–131	48	–76	–5
(II)	0.0	1592	7040	1722	6518	1624	5642	1378	4493
	2/3	–359	–832	–297	–443	–219	–361	–97	–90
	1.0	–210	–604	–222	–481	–170	–464	–56	–300
	NT	–181	–355	–182	–183	–111	–214	–9	–91
(III)	0.0	–264	–141	–26	915	210	1726	529	2615
	2/3	–536	–2232	–455	–1760	–415	–1401	–280	–927
	1.0	923	564	541	270	383	185	273	227
	NT	50	–531	89	–328	101	–127	126	72

X^2 statistic with those for the NT statistic, almost 90 percent of the values of the NT statistic are less than those for Pearson's X^2 statistic. Therefore, the NT statistic often performs better than Pearson's X^2 statistic in a 2-way contingency table.

Furthermore, we consider the real data given by Table 9.6. Table 9.6 shows multiple sclerosis diagnoses for patients in New Orleans by a neurologist in Winnipeg and

Table 9.3 Values of $I(\alpha) \times 10^5$ for 6×6 tables in Cases (I)–(III) for statistics R_M^a ($a = 0.0, 2/3, 1.0$) and $C_{\phi_{NT}}$ when $\alpha = 0.01$ and 0.05 and sparseness index $SI = 2.0, 2.5, 3.0$, and 4.0

6×6		$SI = 2.0$		$SI = 2.5$		$SI = 3.0$		$SI = 4.0$	
Case	$a \backslash \alpha$	0.01	0.05	0.01	0.05	0.01	0.05	0.01	0.05
(I)	0.0	2929	10335	2690	8963	2345	7288	1613	4897
	2/3	–324	–533	–188	–220	–162	21	–43	–46
	1.0	–240	–537	–160	–410	–163	–231	–76	–260
	NT	171	–122	–78	–25	–95	60	–17	–60
(II)	0.0	2316	8820	2317	8416	2011	7179	1593	5251
	2/3	–361	–816	–281	–374	–254	–354	–172	–147
	1.0	–161	–528	–170	–340	–179	–421	–143	–281
	NT	–130	–208	–118	–35	–125	–150	–90	–61
(III)	0.0	–328	–276	0	1136	216	2020	596	3137
	2/3	–617	–2468	–514	–1906	–439	–1607	–355	–1101
	1.0	961	862	675	682	514	452	344	309
	NT	66	–439	154	–79	183	–50	163	79

Table 9.4 A 4×5 table from Simonoff (2003)

	Number in top ten				
Region	0	1	2	3	4
Northeast	3	4	6	2	2
South	8	13	6	1	0
Midwest	2	3	5	2	1
West	1	5	4	3	1

Table 9.5 Values of $I(\alpha) \times 10^5$ for a model given by the estimated values of real data given by Table 9.4 for statistics R_M^a ($a = 0.0, 2/3, 1.0$) and $C_{\phi_{NT}}$ when $\alpha = 0.01$ and 0.05

$SI = 3.6$	a	0.01	0.05
	0.0	827	3401
	2/3	-309	-795
	1.0	-166	-581
	NT	-162	-471

a neurologist in New Orleans. The diagnostic classes are (1) certain; (2) probable; (3) possible; and (4) doubtful, unlikely, or definitely not. The data were obtained by Landis and Koch (1977) and are quoted in Agresti (2013). We consider a model by using the estimated marginal probabilities $p_{j\cdot}$, $j = 1, \ldots, 4$ and $p_{\cdot k}$, $k = 1, \ldots, 4$ from the real data given in Table 9.6 as true values. Table 9.7 shows the results of a simulation similar to Case (I)–Case (III) for the model. In this model, Pearson's X^2

Table 9.6 A 4×4 table from Agresti (2013)

New Orleans	Winnipeg Neurologist			
Neurologist	1	2	3	4
1	5	3	0	0
2	3	11	4	0
3	2	13	3	4
4	1	2	4	14

Table 9.7 Values of $I(\alpha) \times 10^5$ for a model given by the estimated values of real data given by Table 9.6 for statistics R_M^a ($a = 0.0, 2/3, 1.0$) and $C_{\phi_{NT}}$ when $\alpha = 0.01$ and 0.05

$SI = 4.31$	a	0.01	0.05
	0.0	912	3584
	2/3	-173	-128
	1.0	-82	-196
	NT	-69	-70

statistic does not perform better than the Cressie-Read statistic with $a = 2/3$ ($R_M^{2/3}$); however, the NT statistic performs better than $R_M^{2/3}$.

Next, we consider 3-way and 4-way contingency tables. For a 3-way $J \times K \times L$ contingency table, the following sets of marginal probabilities are considered.

Case (I): All marginal probabilities are equal, that is,

$$p_{j\cdot\cdot} = J^{-1},\ j = 1, \ldots, J,\quad p_{\cdot k\cdot} = K^{-1},\ k = 1, \ldots, K,\quad p_{\cdot\cdot\ell} = L^{-1},\ \ell = 1, \ldots, L.$$

Case (II): Marginal probabilities $p_{j\cdot\cdot}$, $p_{\cdot k\cdot}$, and $p_{\cdot\cdot\ell}$ increase linearly corresponding to j, k, and ℓ, that is,

$$p_{j\cdot\cdot} = \frac{1 + J + 2j}{2J(J+1)},\quad j = 1, \ldots, J,\quad p_{\cdot k\cdot} = \frac{1 + K + 2k}{2K(K+1)},\quad k = 1, \ldots, K,$$

$$p_{\cdot\cdot\ell} = \frac{1 + L + 2\ell}{2L(L+1)},\quad \ell = 1, \ldots, L.$$

Case (III): Marginal probabilities $p_{j\cdot\cdot}$, $p_{\cdot k\cdot}$, and $p_{\cdot\cdot\ell}$ are similar to those used for the investigation of a sparse contingency table by Koehler (1986), that is,

$$p_{j\cdot\cdot} = J^{-1}\left(0.1 + 0.9\sum_{m=j}^{J} m^{-1}\right),\quad j = 1, \ldots, J,$$

$$p_{\cdot k \cdot} = K^{-1}\left(0.1 + 0.9 \sum_{m=k}^{K} m^{-1}\right), \quad k = 1, \ldots, K,$$

$$p_{\cdot \cdot \ell} = L^{-1}\left(0.1 + 0.9 \sum_{m=\ell}^{L} m^{-1}\right), \quad \ell = 1, \ldots, L.$$

Similar to 2-way and 3-way tables, for a 4-way $J \times K \times L \times M$ contingency table, we consider

Case (I): All marginal probabilities are equal.

Case (II): Marginal probabilities $p_{j\cdots}$, $p_{\cdot k \cdot\cdot}$, $p_{\cdot\cdot\ell\cdot}$, and $p_{\cdots m}$ increase linearly corresponding to j, k, ℓ, and m.

Case (III): Marginal probabilities $p_{j\cdots}$, $p_{\cdot k \cdot\cdot}$, $p_{\cdot\cdot\ell\cdot}$, and $p_{\cdots m}$ are similar to those used for the investigation of a sparse contingency table by Koehler (1986).

The values of $I(\alpha) \times 10^5$ for R_M^a $(a = 0, 2/3, 1)$ and $C_{\phi_{NT}}$, provided that $N_1 = 10^6$ when the level of significance $\alpha = 0.01, 0.05$ and index $SI = 2.0 \sim 4.0$ for $2 \times 3 \times 4$, $3 \times 3 \times 3$, $2 \times 2 \times 2 \times 2$, and $2 \times 2 \times 3 \times 3$ contingency tables are shown in Tables 9.8, 9.9, 9.10, and 9.11, respectively.

From Tables 9.8, 9.9, 9.10, and 9.11, by seeing the value of $|I(\alpha)|$, we also find that the chi-square approximation of Pearson's X^2 statistic and the NT statistic performs well even when the sample size is not so large in 3-way and 4-way contingency tables.

Next, we consider a comparison between Pearson's X^2 statistic and the NT statistic. From Tables 9.8, 9.9, 9.10, and 9.11, when we compare the values of $|I(\alpha)|$ for Pearson's X^2 statistic with those for the NT statistic, the values of the NT statistic

Table 9.8 Values of $I(\alpha) \times 10^5$ for $2 \times 3 \times 4$ tables in Cases (I)–(III) for statistics R_M^a $(a = 0.0, 2/3, 1.0)$ and $C_{\phi_{NT}}$ when $\alpha = 0.01$ and 0.05 and sparseness index $SI = 2.0, 2.5, 3.0$, and 4.0

$2 \times 3 \times 4$		$SI = 2.0$		$SI = 2.5$		$SI = 3.0$		$SI = 4.0$	
Case	$a \backslash \alpha$	0.01	0.05	0.01	0.05	0.01	0.05	0.01	0.05
(I)	0.0	1985	7669	1925	6773	1681	5715	1265	4089
	2/3	-271	-562	-190	-357	-175	-252	-87	-79
	1.0	-139	-460	-116	-384	-141	-370	-62	-199
	NT	-71	-145	-60	-106	-99	-131	-14	-9
(II)	0.0	1477	6463	1658	6321	1554	5668	1284	4317
	2/3	-381	-923	-251	-514	-195	-353	-170	-150
	1.0	-117	-563	-73	-396	-84	-323	-105	-240
	NT	-97	-317	-40	-166	-51	-71	-57	-50
(III)	0.0	-354	-359	-97	693	92	1295	382	2277
	2/3	-410	-1829	-384	-1596	-319	-1290	-204	-878
	1.0	961	1118	708	730	579	562	470	509
	NT	389	285	328	189	307	227	305	347

Table 9.9 Values of $I(\alpha) \times 10^5$ for $3 \times 3 \times 3$ tables in Cases (I)–(III) for statistics R_M^a ($a = 0.0, 2/3, 1.0$) and $C_{\phi NT}$ when $\alpha = 0.01$ and 0.05 and sparseness index $SI = 2.0, 2.52, 3.0$, and 4.0

$3 \times 3 \times 3$		$SI = 2.0$		$SI = 2.52$		$SI = 3.0$		$SI = 4.0$	
Case	$a \backslash \alpha$	0.01	0.05	0.01	0.05	0.01	0.05	0.01	0.05
(I)	0.0	2082	8285	2012	7159	1768	6037	1307	4245
	2/3	-268	-622	-255	-393	-189	-267	-90	-163
	1.0	-100	-442	-166	-430	-139	-365	-82	-293
	NT	-47	-123	-99	-100	-88	-85	-24	-103
(II)	0.0	1583	7024	1715	6578	1644	5866	1374	4602
	2/3	-399	-901	-284	-564	-186	-364	-115	-136
	1.0	-127	-440	-110	-342	-65	-261	-45	-114
	NT	-99	-184	-67	-63	-13	-7	0	62
(III)	0.0	-333	-571	-72	567	57	1377	379	2317
	2/3	-400	-1931	-377	-1551	-316	-1354	-225	-912
	1.0	1090	1244	811	964	639	737	462	585
	NT	471	340	386	369	355	375	292	396

Table 9.10 Values of $I(\alpha) \times 10^5$ for $2 \times 2 \times 2 \times 2$ tables in Cases (I)–(III) for statistics R_M^a ($a = 0.0, 2/3, 1.0$) and $C_{\phi NT}$ when $\alpha = 0.01$ and 0.05 and sparseness index $SI = 2.0, 2.5, 3.0$, and 4.0

$2 \times 2 \times 2 \times 2$		$SI = 2.0$		$SI = 2.5$		$SI = 3.0$		$SI = 4.0$	
Case	$a \backslash \alpha$	0.01	0.05	0.01	0.05	0.01	0.05	0.01	0.05
(I)	0.0	1403	6028	1439	5522	1348	4814	1074	3318
	2/3	-294	-506	-195	-333	-191	-172	-73	-59
	1.0	-144	-388	-98	-364	-130	-291	-60	-197
	NT	-111	-130	-43	-113	-74	-42	-6	-47
(II)	0.0	1060	5153	1139	4871	1235	4539	978	3552
	2/3	-371	-715	-221	-594	-121	-276	-96	-196
	1.0	-113	-351	-77	-412	-10	-259	-13	-242
	NT	-104	-172	-43	-191	43	-56	11	-86
(III)	0.0	-299	-161	-101	638	46	1243	294	1822
	2/3	-317	-1552	-313	-1275	-256	-1058	-213	-821
	1.0	869	753	605	514	481	407	316	179
	NT	472	210	359	195	285	237	218	130

are less than those for Pearson's X^2 statistic with only one exception (Case (II), $SI = 3.0$, and $\alpha = 0.01$ in Table 9.10). The NT statistic is therefore considered to perform much better than Pearson's X^2 statistic in 3-way and 4-way contingency tables when the sample sizes is not so large.

Further comparisons between the NT statistic and Pearson's X^2 statistic were shown in Taneichi and Sekiya (2025).

Table 9.11 Values of $I(\alpha) \times 10^5$ for $2 \times 2 \times 3 \times 3$ tables in Cases (I)–(III) for statistics R_M^a ($a = 0.0, 2/3, 1.0$) and $C_{\phi NT}$ when $\alpha = 0.01$ and 0.05 and sparseness index $SI = 2.0, 2.5, 3.0$, and 4.0

$2 \times 2 \times 3 \times 3$		$SI = 2.0$		$SI = 2.5$		$SI = 3.0$		$SI = 4.0$	
Case	$a \backslash \alpha$	0.01	0.05	0.01	0.05	0.01	0.05	0.01	0.05
(I)	0.0	2713	9804	2462	8449	2181	6824	1463	4675
	2/3	-334	-647	-198	-434	-194	-299	-119	-191
	1.0	-150	-333	-66	-268	-73	-271	-42	-224
	NT	-82	41	-2	69	-24	37	1	-33
(II)	0.0	1981	8235	2116	7658	1994	6776	1506	5168
	2/3	-332	-1017	-316	-630	-241	-481	-140	-274
	1.0	-46	-347	-85	-274	-90	-254	-25	-135
	NT	-22	-90	-45	-1	-53	16	10	50
(III)	0.0	-446	-972	-220	285	12	1163	383	2355
	2/3	-446	-2133	-390	-1806	-360	-1531	-272	-1204
	1.0	1346	1630	996	1322	803	1003	635	787
	NT	595	523	510	561	428	487	398	547

9.3 Effect of a Test by the NT Statistic

In Sect. 9.2, we find that performance for the approximation of Pearson's X^2 statistic and the NT statistic to a chi-square distribution is good even when the sample size is not so large, and the NT statistic performs better than Pearson's X^2 statistic. In this section, we show how the performance of the approximation of the test statistics affects the results of a practical test. A 4×4 artificial contingency table (Table A) with $SI = 2.0$ is derived from Table 9.12. We test independent hypothesis $H_0^{(I)}$ of contingency table (Table A) based on Pearson's X^2 statistic and the NT statistic by using the nominal critical point $\chi_9^2(0.05) \approx 16.919$. The observed values of Pearson's X^2 statistic and the NT statistic are 16.857 and 17.057, respectively. Therefore, $H_0^{(I)}$ is accepted on the basis of Pearson's X^2 statistic, and $H_0^{(I)}$ is rejected on the basis of the NT statistic at a significance level of 0.05. On the other hand, the exact p-value derived using Monte Carlo simulation, which we call the simulated p-value of Pearson's X^2 statistic and the NT statistic, are 0.0427 and 0.0401, respectively. That is, if we test more exactly, we conclude that $H_0^{(I)}$ is rejected by both Pearson's X^2 statistic and the NT statistic with a significance level of 0.05.

Table 9.12 An artificial 4×4 contingency table A with $SI = 2.0$

1	0	2	1
0	1	1	5
2	6	0	2
1	3	5	2

Table 9.13 An artificial 4 × 4 contingency table B with $SI = 2.0$

12	3	4	1
2	2	2	0
3	1	1	0
0	0	0	1

Table 9.14 Conclusions of tests by using the NT statistic and Pearson's X^2 statistic for which distributions are approximated by a chi-square distribution and a simulated exact distribution, respectively

	X^2 statistic		NT statistic	
Table	Chi-square approx.	Exact approx.	Chi-square approx.	Exact approx.
A	Accept	Reject	Reject	Reject
B	Reject	Accept	Accept	Accept

Next, we consider an artificial 4 × 4 contingency table (Table B) derived from Table 9.13. Observed values of Pearson's X^2 statistic and the NT statistic are 17.631 and 15.240, respectively. Therefore, $H_0^{(I)}$ is rejected on the basis of Pearson's X^2 statistic and accepted on the basis of the NT statistic by using a nominal critical point at a significant level of 0.05. On the other hand, simulated p-values of Pearson's X^2 statistic and the NT statistic are 0.0824 and 0.0577, respectively. That is, if we test more exactly, we conclude that $H_0^{(I)}$ is accepted on the basis of both Pearson's X^2 statistic and the NT statistic with a significance level of 0.05.

For contingency tables (Tables A and B), we find the following. For the NT statistic, the conclusion of the tests using a nominal significance level agree with those of a test based on the exact distribution of the statistic. In contrast, for Pearson's X^2 statistic, the conclusions of the tests using a nominal critical point is opposite to those of a test based on the exact distribution of the statistic. Results of the tests are summarized in Table 9.14.

These results suggest that the NT provides a better approximation of the chi-square distribution compared to Pearson's X^2.

References

Agresti A (2013) Categorical data analysis, 3rd edn. Wiley, New York

Abzug R, Simonoff JS, Ahlstrom D (2000) Nonprofits as large employers: A city-level geographical inquiry. Nonprofit and Voluntary Sector Quarterly 29:455–470

Koehler KJ (1986) Goodness-of-fit tests for log-linear models in sparse contingency tables. J Am Stat Assoc 81:483–493

Landis JR, Koch GG (1977) An application of hierarchical kappa-type statistics in the assesment of majority agreement among multiple observers. Biometrics 33:363–374

Rukhin AL (1994) Optimal estimator for the mixture parameter by the method of moments and information affinity. in Trans. 12th Prague Conference on Information Theory, p 214–219

Simonoff JS (2003) Analyzing Categorical Data. Springer-Verlag, New York

Taneichi N, Sekiya Y (2025) A new test statistic based on ϕ-divergence for testing independence of contingency tables that minimizes the second-order correction term (in Japanese). J Jpn Statist Soc Ser J 55(1):1–24

Chapter 10
Summary and Conclusion

Keywords Complete independence · Conditional independence · Contingency table · Generalized linear model · Independence among groups of factors · Loglinear model · Second-order correction term

10.1 Part One

We consider tests of independence of 2-way contingency tables, tests of complete independence of multi-way contingency tables, tests of independence among groups of factors, and tests of conditional independence of 3-way contingency tables. We also consider goodness-of-fit tests for a generalized linear model of binary data. On the basis of the continuous term of an asymptotic expansion, an approximation of the distribution of the test statistic based on ϕ-divergence was derived. Using the expressions, we obtained the transformed statistics that improve the speed of convergence to a chi-square limiting distribution. As a special case of ϕ-divergence statistics, Taneichi and Sekiya (2007), Taneichi et al. (2014, 2019, 2021) numerically compared the original power divergence test statistics R^a $(a = 0, 0.2, 2/3, 1)$ with the transformed test statistics for the performance of speed of convergence to a chi-square distribution and power. These comparisons found that the performance of the transformed test statistics for speed of convergence to a chi-square distribution was usually much better than the original test statistics. Particularly, the transformed statistics of $R^0 \equiv G^2$ (the log likelihood ratio statistic) performed extremely well, and the transformed statistic of R^a with $a = 0.2$ performed better than that of R^a with $a = 2/3$ and $a = 1$. For the comparison of the power of statistics, the power of the transformed statistics was almost the same as the power of the original statistics. Furthermore, on the basis of an asymptotic expansion, an approximation of the distribution of the log likelihood ratio test statistic G^2 for a loglinear model of a multi-way contingency table was derived. Using this expression, we obtained the Bartlett-type

N. Taneichi and Y. Sekiya, *Improving Tests for Discrete Small Sample Data*,
JSS Research Series in Statistics,
https://doi.org/10.1007/978-981-95-5301-3_10

transformed statistic G_B^2 that improves the speed of convergence to a chi-square limiting distribution. A numerical comparison showed that the transformed statistic G_B^2 is effective for improving the speed of convergence. The numerical comparison also showed that the power of statistic G_B^2 is almost the same as that of the original statistic G^2.

This improvement increases the reliability of the results of asymptotic tests.

10.2 Part Two

For the test of complete independence of a multi-way contingency table based on the ϕ-divergence family, by using a second-order correction term, we derived conditions (9.1) for selecting a ϕ-divergence statistic when considering an asymptotic test for cases in which the sample size is not so large. Concretely, only Pearson's X^2 statistic satisfies conditions (9.1) among the the power divergence family of statistics as well as among the family of statistics proposed by Rukhin (1994). We present a new test statistic (NT statistic) that satisfies the conditions (9.1) as does Pearson's X^2 statistic. Furthermore, by a numerical comparison, it is found that the distribution of the NT statistic is well approximated by a chi-square distribution, as is the case with Pearson's X^2 statistic. With regard to comparing Pearson's X^2 statistic and the NT statistic, the number of cases in which the NT statistic outperforms Pearson's X^2 statistic is much more than the other way around. This result is more conspicuous in 3-way and 4-way contingency tables.

In this way, in the case of tests of complete independence and independence among groups of factors in a multi-way contingency table, the NT statistic is considered to be effective when the sample size is not so large.

References

Rukhin AL (1994) Optimal estimator for the mixture parameter by the method of moments and information affinity. In: Trans. 12th Prague conference on information theory, pp 214–219

Taneichi N, Sekiya Y (2007) Improved transformed statistics for the test of independence in $r \times s$ contingency tables. J Multivariate Anal 98:1630–1657

Taneichi N, Sekiya Y, Toyama J (2014) Transformed goodness-of-fit statistics for a generalized linear model of binary data. J Multivariate Anal 123:311–329

Taneichi N, Sekiya Y, Toyama J (2019) Transformed statistics for tests of conditional independence in $J \times K \times L$ contingency tables. J Multivariate Anal 171:193–208

Taneichi N, Sekiya Y, Toyama J (2021) Improvement of the test of independence among groups of factors in a multi-way contingency table. Jpn J Stat Data Sci 4:181–213

Appendix A
Proofs of Important Results

A.1 Proof of Theorem 3.2

By the transformation (3.5), statistic $C_\phi^{[2]}$ can be rewritten as

$$C_\phi^{[2]}(U) = 2n \sum_{j=1}^{J} \sum_{k=1}^{K} p_{j\cdot} p_{\cdot k} \left(1 + \frac{U_{j\cdot}}{\sqrt{n} p_{j\cdot}}\right) \left(1 + \frac{U_{\cdot k}}{\sqrt{n} p_{\cdot k}}\right) \phi(D^*), \qquad \text{(A.1)}$$

where

$$D^* = \left(1 + \frac{U_{j\cdot}}{\sqrt{n} p_{j\cdot}}\right)^{-1} \left(1 + \frac{U_{\cdot k}}{\sqrt{n} p_{\cdot k}}\right)^{-1} \left(1 + \frac{U_{jk}}{\sqrt{n} p_{j\cdot} p_{\cdot k}}\right),$$

$U_{j\cdot} = \sum_{k=1}^{K} U_{jk}$, $j = 1, \ldots, J$ and $U_{\cdot k} = \sum_{j=1}^{J} U_{jk}$, $k = 1, \ldots, K$. If we regard

$$f(u) \left\{1 + n^{-1/2} h_1(u) + n^{-1} h_2(u)\right\},$$

where f, h_1 and h_2 are given by (3.7), (3.9) and (3.10), respectively, as the continuous density function of U, then we can regard

$$J_1^* = \int \cdots \int_{B_\phi(b)} f(u) \left\{1 + n^{-1/2} h_1(u) + n^{-1} h_2(u)\right\} du$$

as the distribution function of $C_\phi^{[2]}(U)$, where $du = du_{11} \cdots du_{J,K-1}$. So, the characteristic function of $C_\phi^{[2]}(U)$ is calculated as

$$c_\phi(t) = \int_{-\infty}^{\infty} \cdots \int_{-\infty}^{\infty} \exp\left\{it C_\phi^{[2]}(u)\right\} f(u) \left\{1 + n^{-1/2} h_1(u) + n^{-1} h_2(u)\right\} du.$$

N. Taneichi and Y. Sekiya, *Improving Tests for Discrete Small Sample Data*, JSS Research Series in Statistics,
https://doi.org/10.1007/978-981-95-5301-3

We can expand $C_\phi^{[2]}(u)$ as

$$C_\phi^{[2]}(u) = K_\phi(u) + o(n^{-1}), \tag{A.2}$$

where

$$\begin{aligned}
K_\phi(u) = &\left(\sum_{j=1}^{J}\sum_{k=1}^{K} \frac{u_{jk}^2}{p_{j\cdot}p_{\cdot k}} - \sum_{j=1}^{J} \frac{u_{j\cdot}^2}{p_{j\cdot}} - \sum_{k=1}^{K} \frac{u_{\cdot k}^2}{p_{\cdot k}} \right) \\
&+ n^{-1/2}\left[\frac{1}{3}\phi'''(1) \sum_{j=1}^{J}\sum_{k=1}^{K} \frac{u_{jk}^3}{p_{j\cdot}^2 p_{\cdot k}^2} + 2\left\{1 + \phi'''(1)\right\} \sum_{j=1}^{J}\sum_{k=1}^{K} \frac{u_{j\cdot}u_{\cdot k}u_{jk}}{p_{j\cdot}p_{\cdot k}} \right. \\
&\quad + \left\{1 + \frac{2}{3}\phi'''(1)\right\} \left(\sum_{j=1}^{J} \frac{u_{j\cdot}^3}{p_{j\cdot}^2} + \sum_{k=1}^{K} \frac{u_{\cdot k}^3}{p_{\cdot k}^2} \right) \\
&\quad \left. - \left\{1 + \phi'''(1)\right\} \left(\sum_{j=1}^{J}\sum_{k=1}^{K} \frac{u_{j\cdot}u_{jk}^2}{p_{j\cdot}^2 p_{\cdot k}} + \sum_{j=1}^{J}\sum_{k=1}^{K} \frac{u_{\cdot k}u_{jk}^2}{p_{j\cdot}p_{\cdot k}^2} \right) \right] \\
&+ n^{-1}\left[\frac{1}{12}\phi^{(4)}(1) \sum_{j=1}^{J}\sum_{k=1}^{K} \frac{u_{jk}^4}{p_{j\cdot}^3 p_{\cdot k}^3} + \left\{1 + 2\phi'''(1) + \frac{1}{2}\phi^{(4)}(1)\right\} \sum_{j=1}^{J}\sum_{k=1}^{K} \frac{u_{j\cdot}^2 u_{\cdot k}^2}{p_{j\cdot}p_{\cdot k}} \right. \\
&\quad - \left\{1 + \frac{4}{3}\phi'''(1) + \frac{1}{4}\phi^{(4)}(1)\right\} \left(\sum_{j=1}^{J} \frac{u_{j\cdot}^4}{p_{j\cdot}^3} + \sum_{k=1}^{K} \frac{u_{\cdot k}^4}{p_{\cdot k}^3} \right) \\
&\quad - \frac{1}{3}\left\{2\phi'''(1) + \phi^{(4)}(1)\right\} \left(\sum_{j=1}^{J}\sum_{k=1}^{K} \frac{u_{j\cdot}u_{jk}^3}{p_{j\cdot}^3 p_{\cdot k}^2} + \sum_{j=1}^{J}\sum_{k=1}^{K} \frac{u_{\cdot k}u_{jk}^3}{p_{j\cdot}^2 p_{\cdot k}^3} \right) \\
&\quad + \left\{1 + 2\phi'''(1) + \frac{1}{2}\phi^{(4)}(1)\right\} \left(\sum_{j=1}^{J}\sum_{k=1}^{K} \frac{u_{j\cdot}^2 u_{jk}^2}{p_{j\cdot}^3 p_{\cdot k}} + \sum_{j=1}^{J}\sum_{k=1}^{K} \frac{u_{\cdot k}^2 u_{jk}^2}{p_{j\cdot}p_{\cdot k}^3} \right) \\
&\quad + \left\{1 + 3\phi'''(1) + \phi^{(4)}(1)\right\} \sum_{j=1}^{J}\sum_{k=1}^{K} \frac{u_{j\cdot}u_{\cdot k}u_{jk}^2}{p_{j\cdot}^2 p_{\cdot k}^2} \\
&\quad - \left\{2 + 4\phi'''(1) + \phi^{(4)}(1)\right\} \\
&\quad \left. \times \left(\sum_{j=1}^{K}\sum_{k=1}^{K} \frac{u_{j\cdot}^2 u_{\cdot k}u_{jk}}{p_{j\cdot}^2 p_{\cdot k}} + \sum_{j=1}^{J}\sum_{k=1}^{K} \frac{u_{j\cdot}u_{\cdot k}^2 u_{jk}}{p_{j\cdot}p_{\cdot k}^2} \right) \right].
\end{aligned} \tag{A.3}$$

Since

$$\exp\left\{itC_\phi(u)\right\} = \left[\exp\left\{it\left(\sum_{j=1}^{J}\sum_{k=1}^{K}\frac{u_{jk}^2}{p_{j\cdot}p_{\cdot k}} - \sum_{j=1}^{J}\frac{u_{j\cdot}^2}{p_{j\cdot}} - \sum_{k=1}^{K}\frac{u_{\cdot k}^2}{p_{\cdot k}}\right)\right\}\right] \times\left[1 + n^{-1/2}g_1^\phi(u) + n^{-1}g_2^\phi(u) + o(n^{-1})\right],$$

where

$$\begin{aligned} g_1^\phi(u) = it\Bigg[& \frac{1}{3}\phi'''(1)\sum_{j=1}^{J}\sum_{k=1}^{K}\frac{u_{jk}^3}{p_{j\cdot}^2 p_{\cdot k}^2} + 2\left\{1+\phi'''(1)\right\}\sum_{j=1}^{J}\sum_{k=1}^{K}\frac{u_{j\cdot}u_{\cdot k}u_{jk}}{p_{j\cdot}p_{\cdot k}} \\ & + \left\{1 + \frac{2}{3}\phi'''(1)\right\}\left(\sum_{j=1}^{J}\frac{u_{j\cdot}^3}{p_{j\cdot}^2} + \sum_{k=1}^{K}\frac{u_{\cdot k}^3}{p_{\cdot k}^2}\right) \\ & - \left\{1+\phi'''(1)\right\}\left(\sum_{j=1}^{J}\sum_{k=1}^{K}\frac{u_{j\cdot}u_{jk}^2}{p_{j\cdot}^2 p_{\cdot k}} + \sum_{j=1}^{J}\sum_{k=1}^{K}\frac{u_{\cdot k}u_{jk}^2}{p_{j\cdot}p_{\cdot k}^2}\right)\Bigg] \end{aligned}$$

and

$$\begin{aligned} g_2^\phi(u) = & \frac{1}{2}\left\{g_1^\phi(u)\right\}^2 \\ & + it\Bigg[\frac{1}{12}\phi^{(4)}(1)\sum_{j=1}^{J}\sum_{k=1}^{K}\frac{u_{jk}^4}{p_{j\cdot}^3 p_{\cdot k}^3} + \left\{1 + 2\phi'''(1) + \frac{1}{2}\phi^{(4)}(1)\right\}\sum_{j=1}^{J}\sum_{k=1}^{K}\frac{u_{j\cdot}^2 u_{\cdot k}^2}{p_{j\cdot}p_{\cdot k}} \\ & - \left\{1 + \frac{4}{3}\phi'''(1) + \frac{1}{4}\phi^{(4)}(1)\right\}\left(\sum_{j=1}^{J}\frac{u_{j\cdot}^4}{p_{j\cdot}^3} + \sum_{k=1}^{K}\frac{u_{\cdot k}^4}{p_{\cdot k}^3}\right) \\ & - \frac{1}{3}\left\{2\phi'''(1) + \phi^{(4)}(1)\right\}\left(\sum_{j=1}^{J}\sum_{k=1}^{K}\frac{u_{j\cdot}u_{jk}^3}{p_{j\cdot}^3 p_{\cdot k}^2} + \sum_{j=1}^{J}\sum_{k=1}^{K}\frac{u_{\cdot k}u_{jk}^3}{p_{j\cdot}^2 p_{\cdot k}^3}\right) \\ & + \left\{1 + 2\phi'''(1) + \frac{1}{2}\phi^{(4)}(1)\right\}\left(\sum_{j=1}^{J}\sum_{k=1}^{K}\frac{u_{j\cdot}^2 u_{jk}^2}{p_{j\cdot}^3 p_{\cdot k}} + \sum_{j=1}^{J}\sum_{k=1}^{K}\frac{u_{\cdot k}^2 u_{jk}^2}{p_{j\cdot}p_{\cdot k}^3}\right) \\ & + \left\{1 + 3\phi'''(1) + \phi^{(4)}(1)\right\}\sum_{j=1}^{J}\sum_{k=1}^{K}\frac{u_{j\cdot}u_{\cdot k}u_{jk}^2}{p_{j\cdot}^2 p_{\cdot k}^2} \\ & - \left\{2 + 4\phi'''(1) + \phi^{(4)}(1)\right\}\left(\sum_{j=1}^{J}\sum_{k=1}^{K}\frac{u_{j\cdot}^2 u_{\cdot k}u_{jk}}{p_{j\cdot}^2 p_{\cdot k}} + \sum_{j=1}^{J}\sum_{k=1}^{K}\frac{u_{j\cdot}u_{\cdot k}^2 u_{jk}}{p_{j\cdot}p_{\cdot k}^2}\right)\Bigg], \end{aligned}$$

$c_\phi(t)$ is represented as

$$c'_\phi(t) = \int_{-\infty}^{\infty} \cdots \int_{-\infty}^{\infty} (2\pi)^{-(JK-1)/2} |\Omega|^{-1/2} \\ \times \left[\exp\left\{ -\frac{1}{2} u^\top \Omega^{-1} u + it \left(\sum_{j=1}^{J} \sum_{k=1}^{K} \frac{u_{jk}^2}{p_{j\cdot} p_{\cdot k}} - \sum_{j=1}^{J} \frac{u_{j\cdot}^2}{p_{j\cdot}} - \sum_{k=1}^{K} \frac{u_{\cdot k}^2}{p_{\cdot k}} \right) \right\} \right] \\ \times \left[1 + n^{-1/2} g_1^\phi(u) + n^{-1} g_2^\phi(u) \right] \left[1 + n^{-1/2} h_1(u) + n^{-1} h_2(u) \right] du + o(n^{-1}).$$

Furthermore, if we put matrix Λ as

$$\Lambda = \Lambda_1 + \Lambda_2, \tag{A.4}$$

where

$$\Lambda_1 = (I_{JK-1}\ O_{JK-1})\, A_1\, (I_{JK-1}\ O_{JK-1})^\top,$$

$$\Lambda_2 = \left(\frac{1}{1-2it} - 1 \right) (I_{JK-1}\ O_{JK-1})\, A_2\, (I_{JK-1}\ O_{JK-1})^\top,$$

I_{JK-1} being the $(JK-1) \times (JK-1)$ identity matrix and O_{JK-1} being the $(JK-1)$ dimensional zero vector,

$$A_1 = \mathrm{diag}(p_{1\cdot}, \ldots, p_{J\cdot}) \otimes \mathrm{diag}(p_{\cdot 1}, \ldots, p_{\cdot K}) - (p_L p_L^\top) \otimes (p_C p_C^\top),$$

$$A_2 = \left(\mathrm{diag}(p_{1\cdot}, \ldots, p_{J\cdot}) - p_L p_L^\top \right) \otimes \left(\mathrm{diag}(p_{\cdot 1}, \ldots, p_{\cdot K}) - p_C p_C^\top \right),$$

$p_L = (p_{1\cdot}, \ldots, p_{J\cdot})^\top$, $p_C = (p_{\cdot 1}, \ldots, p_{\cdot K})^\top$, and $\otimes$ denotes the Kronecker product, then

$$c_\phi(t) = \left(\frac{|\Omega|}{|\Lambda|} \right)^{-1/2} \int_{-\infty}^{\infty} \cdots \int_{-\infty}^{\infty} (2\pi)^{-(JK-1)/2} |\Lambda|^{-1/2} \\ \times \left\{ \exp\left(-\frac{1}{2} u^\top \Lambda^{-1} u \right) \right\} A_\phi(u) du + o(n^{-1}), \tag{A.5}$$

where

$$A_\phi(u) = 1 + n^{-1/2} \left\{ h_1(u) + g_1^\phi(u) \right\} + n^{-1} \left\{ h_2(u) + h_1(u) g_1^\phi(u) + g_2^\phi(u) \right\},$$

and Ω is given by (3.8). By carrying out the integration of (A.5), the characteristic function $c_\phi(t)$ is expanded as

$$c_\phi(t) = \left(\frac{|\Omega|}{|\Lambda|} \right)^{-1/2} \left[1 + n^{-1} \sum_{\ell=0}^{3} (1-2it)^{-\ell} w_\ell^\phi + o(n^{-1}) \right], \tag{A.6}$$

where w_ℓ^ϕ, $\ell = 0, 1, 2, 3$ are defined in (3.14). Since

$$\left(\frac{|\Omega|}{|\Lambda|}\right)^{-1/2} = (1 - 2it)^{-(J-1)(K-1)/2}, \tag{A.7}$$

then $(|\Omega|/|\Lambda|)^{-1/2}$ is the characteristic function of the chi-square distribution with degrees of freedom $(J-1)(K-1)$. Therefore, by inverting (A.6), we obtain (3.14). We have completed the proof of Theorem 3.2.

A.2 Proof of Theorem 3.3

The function $S_1(\sqrt{n}u_\ell^* + np_\ell^*)$ defined by (3.21) is differentiable except when $u_\ell^* \in L_\ell$ defined by (3.19), and $f(u^*)$ is a differentiable function on R_{JK-1}. Therefore, by (8.10) in the proof of Lemma 1 of Yarnold (1972) cited in Chap. 3,

$$[S_1(\sqrt{n}u_\ell^* + np_\ell^*)f(u^*)]_{\eta_\ell(\tilde{u}_\ell^*)}^{\theta_\ell(\tilde{u}_\ell^*)} = \int_{\eta_\ell(\tilde{u}_\ell^*)}^{\theta_\ell(\tilde{u}_\ell^*)} D_\ell S_1(\sqrt{n}u_\ell^* + np_\ell^*)f(u^*)du_\ell^*$$

$$+ \sum_{\substack{u_\ell^* = \eta_\ell(\tilde{u}_\ell^*) \\ u_\ell^* \in L_\ell}}^{\theta_\ell(\tilde{u}_\ell^*)} \Delta_\ell S_1(\sqrt{n}u_\ell^* + np_\ell^*)f(u^*), \tag{A.8}$$

where

$$[h(u^*)]_{\eta_\ell(\tilde{u}_\ell^*)}^{\theta_\ell(\tilde{u}_\ell^*)} = h(u_1^*, \ldots, u_{\ell-1}^*, \theta_\ell(\tilde{u}_\ell^*), u_{\ell+1}^*, \ldots, u_{JK-1}^*)$$

$$- h(u_1^*, \ldots, u_{\ell-1}^*, \eta_\ell(\tilde{u}_\ell^*), u_{\ell+1}^*, \ldots, u_{JK-1}^*),$$

$$\begin{aligned} \Delta_\ell F(x) = {} & F(x_1, \ldots, x_{\ell-1}, x_\ell + 0, x_{\ell+1}, \ldots, x_{JK-1}) \\ & - F(x_1, \ldots, x_{\ell-1}, x_\ell - 0, x_{\ell+1}, \ldots, x_{JK-1}), \end{aligned}$$

and $D_\ell F(x) = (\partial/\partial x_\ell)F(x)$. Since $\Delta_\ell S_1(\sqrt{n}u_\ell^* + np_\ell^*)f(u^*) = -f(u^*)$ and

$$D_\ell S_1(\sqrt{n}u_\ell^* + np_\ell^*)f(u^*) = f(u^*)\left\{\sqrt{n} - S_1(\sqrt{n}u_\ell^* + np_\ell^*)\left(\frac{u_\ell^*}{p_\ell^*} - \frac{u_{JK}^*}{p_{JK}^*}\right)\right\},$$

from (A.8),

$$[S_1(\sqrt{n}u_\ell^* + np_\ell^*)f(u^*)]_{\eta_\ell(\tilde{u}_\ell^*)}^{\theta_\ell(\tilde{u}_\ell^*)} = \sqrt{n}\int_{\eta_\ell(\tilde{u}_\ell^*)}^{\theta_\ell(\tilde{u}_\ell^*)} f(u^*)du_\ell^* - \sum_{\substack{u_\ell^* = \eta_\ell(\tilde{u}_\ell^*) \\ u_\ell^* \in L_\ell}}^{\theta_\ell(\tilde{u}_\ell^*)} f(u^*)$$

$$
- \int_{\eta_\ell(\tilde{u}_\ell^*)}^{\theta_\ell(\tilde{u}_\ell^*)} S_1(\sqrt{n}u_\ell^* + np_\ell^*) f(u^*) \left(\frac{u_\ell^*}{p_\ell^*} - \frac{u_{JK}^*}{p_{JK}^*} \right) du_\ell^*. \tag{A.9}
$$

Using (A.9), J_2^* given by (3.18) is represented as

$$
J_2^* = A - B^* + C^*,
$$

where

$$
B^* = \int \cdots \int_{B_\phi(b)} f(u^*) du^*
$$

and

$$
C^* = \sum_{\ell=1}^{JK-1} n^{-(JK-\ell)/2} \sum_{u_{\ell+1}^* \in M_{\ell+1}^\phi} \cdots \sum_{u_{JK-1}^* \in M_{JK-1}^\phi} \int \cdots \int_{B_\phi(b)} S_1(\sqrt{n}u_\ell^* + np_\ell^*) \times \left(\frac{u_\ell^*}{p_\ell^*} - \frac{u_{JK}^*}{p_{JK}^*} \right) f(u^*) du_1^* \cdots du_\ell^*.
$$

Furthermore, $B^* = B + O(n^{-1})$ and $C^* = C + O(n^{-1})$. Therefore, we obtain (3.23). We have completed the proof of Theorem 3.3.

A.3 Proof of Theorem 3.4

Because of (3.11) and (3.16),

$$
\frac{u_{JK-1}^*}{p_{JK-1}^*} - \frac{u_{JK}^*}{p_{JK}^*} = \frac{1}{p_{JK}^*} \left(u_1^* + \cdots + u_{JK-2}^* \right) + \left(\frac{1}{p_{JK-1}^*} + \frac{1}{p_{JK}^*} \right) u_{JK-1}^*. \tag{A.10}
$$

If we put

$$
C_\ell^* = \int \cdots \int_{B_\phi(b)} S_1(\sqrt{n}u_{JK-1}^* + np_{JK-1}^*) u_\ell^* f(u^*) du^*, \quad \ell = 1, \ldots, JK-1, \tag{A.11}
$$

then, from (A.10) and (3.24),

$$
C = n^{-1/2} \left\{ \frac{1}{p_{JK}^*} \left(C_1^* + \cdots + C_{JK-2}^* \right) + \left(\frac{1}{p_{JK-1}^*} + \frac{1}{p_{JK}^*} \right) C_{JK-1}^* \right\}. \tag{A.12}
$$

Since $-1/2 \le S_1(\sqrt{n}u_{JK-1}^* + np_{JK-1}^*) < 1/2$ and (A.11),

$$
|C_\ell^*| \le \frac{1}{2} \int \cdots \int_{B_\phi(b)} |u_\ell^*| f(u^*) du^*, \quad \ell = 1, \ldots, JK-1. \tag{A.13}
$$

If we put

$$\epsilon_\ell^* = \int_{-\infty}^{\infty} \cdots \int_{-\infty}^{\infty} |u_\ell^*| f(u^*) du^*, \quad \ell = 1, \ldots, JK - 1,$$

then

$$\int \cdots \int_{B_\phi(b)} |u_\ell^*| f(u^*) du^* = \epsilon_\ell^* \int \cdots \int_{B_\phi(b)} g_\ell(u^*) du^*, \quad \ell = 1, \ldots, JK - 1, \tag{A.14}$$

where $g_\ell(u^*) = |u_\ell^*| f(u^*) \epsilon_\ell^{*-1}$ and $\int_{-\infty}^{\infty} \cdots \int_{-\infty}^{\infty} g_\ell(u^*) du^* = 1$. If we regard $g_\ell(u^*)$ as the density function of $U^* = (U_1^*, \ldots, U_{JK-1}^*)^\top$, where $U_\ell^* = U_{mk}$, $\ell = (m-1)K + k$ for $m = 1, \ldots, J$, $k = 1, \ldots, K$, then we can regard

$$\int \cdots \int_{B_\phi(b)} g_\ell(u^*) du^*$$

as the distribution function of $C_\phi^{[2]}(U^*)$. By using the relations (3.16) and (3.20), the characteristic function of $C_\phi^{[2]}(U^*)$ is calculated as

$$\begin{aligned} c_{g_\ell}(t) &= \int_{-\infty}^{\infty} \cdots \int_{-\infty}^{\infty} \exp\{it C_\phi^{[2]}(u^*)\} g_\ell(u^*) du^* \\ &= \epsilon_\ell^{*-1} \left\{ \left(\frac{|\Omega|}{|\Lambda|} \right)^{-1/2} F_\ell + O\left(n^{-1/2}\right) \right\}, \end{aligned} \tag{A.15}$$

where

$$F_\ell = \int_{-\infty}^{\infty} \cdots \int_{-\infty}^{\infty} (2\pi)^{-(JK-1)/2} |\Lambda|^{-1/2} |u_\ell^*| \left\{ \exp\left(-\frac{1}{2} u^{*\top} \Lambda^{-1} u^* \right) \right\} du^*, \tag{A.16}$$

Λ being given by (A.4) and Ω being given by (3.8). Carrying out the integral (A.16) by noting the relations (3.16) and (3.20), we obtain

$$F_\ell = \sqrt{\frac{2\lambda_{\ell\ell}}{\pi}},$$

where $\lambda_{\ell\ell}$ is the (ℓ, ℓ) element of Λ, that is,

$$\lambda_{\ell\ell} = p_{m\cdot} p_{\cdot k} (p_{m\cdot} + p_{\cdot k} - 2 p_{m\cdot} p_{\cdot k}) \{1 + (1 - 2it)^{-1} \rho(m, k)\}.$$

Therefore, when $|\rho(m, k)| < 1$, F_ℓ can be expanded as

$$F_\ell = \sqrt{\frac{2\eta(m, k)}{\pi}} \sum_{s=0}^{\infty} \tau_{(m,k)}^{(s)} (1 - 2it)^{-s}, \tag{A.17}$$

where $\eta(m,k) = p_{m\cdot}p_{\cdot k}(p_{m\cdot} + p_{\cdot k} - 2p_{m\cdot}p_{\cdot k})$. From (A.7), (A.15), and (A.17),

$$c_{g_\ell}(t) = 2\,\epsilon_\ell^{*-1}\sqrt{\frac{\eta(m,k)}{2\pi}}(1-2it)^{-(J-1)(K-1)/2}\sum_{s=0}^{\infty}\tau_{(m,k)}^{(s)}(1-2it)^{-s} + O\left(n^{-1/2}\right). \tag{A.18}$$

Inverting (A.18), we obtain

$$\int\cdots\int_{B_\phi(b)} g_\ell(u^*)du^* = 2\,\epsilon_\ell^{*-1}\sqrt{\frac{\eta(m,k)}{2\pi}}\sum_{s=0}^{\infty}\tau_{(m,k)}^{(s)}\Pr\{\chi^2_{(J-1)(K-1)+2s} \le b\} + O\left(n^{-1/2}\right). \tag{A.19}$$

Applying (A.19) to (A.14),

$$\int\cdots\int_{B_\phi(b)} |u_\ell^*| f(u^*)du^* = 2\sqrt{\frac{\eta(m,k)}{2\pi}}\sum_{s=0}^{\infty}\tau_{(m,k)}^{(s)}\Pr\{\chi^2_{(J-1)(K-1)+2s} \le b\} + O\left(n^{-1/2}\right). \tag{A.20}$$

If we put

$$C_{mk} = C_\ell^*,\ \ell = (m-1)K + k \text{ for } m = 1,\ldots,J,\ k = 1,\ldots,K, \tag{A.21}$$

then, from (A.13), (A.20) and (A.21),

$$|C_{mk}| \le \sqrt{\frac{\eta(m,k)}{2\pi}}\sum_{s=0}^{\infty}\tau_{(m,k)}^{(s)}\Pr\{\chi^2_{(J-1)(K-1)+2s} \le b\} + O\left(n^{-1/2}\right). \tag{A.22}$$

From (A.12) and (A.21),

$$|C| \le n^{-1/2}\left(\frac{1}{p_{J\cdot}p_{\cdot K}}\sum_{m=1}^{J}\sum_{\substack{k=1\\(m,k)\ne(J,K)}}^{K}|C_{mk}| + \frac{1}{p_{J\cdot}p_{\cdot K-1}}|C_{J,K-1}|\right). \tag{A.23}$$

By applying (A.22) to (A.23), we obtain (3.26). We have completed the proof of Theorem 3.4.

A.4 Proof of Theorem 4.4

We calculate J_1^* given by (4.19) by the following procedure. First, we derive an approximation of the characteristic function of $C_\phi^{(1)[M]}(U)$ assuming that the distribution of U is continuous, that is,

$$\psi_\phi(t) = \int_{-\infty}^{\infty} \cdots \int_{-\infty}^{\infty} \exp\left\{it\, C_\phi^{(1)[M]}(u)\right\} g^{[M]}(u) \Big\{1 + n^{-1/2} h_1^{[M]}(u) + n^{-1} h_2^{[M]}(u) + n^{-3/2} h_3^{[M]}(u)\Big\} du + O(n^{-2}) \tag{A.24}$$

up to order $n^{-3/2}$. Next, we apply an inversion formula to the approximation.

By the assumption of ϕ, we can expand $C_\phi^{(1)[M]}(u)$ as

$$C_\phi^{(1)[M]}(u) = \tau_0(u) + n^{-1/2}\tau_1^\phi(u) + n^{-1}\tau_2^\phi(u) + n^{-3/2}\tau_3^\phi(u) + O(n^{-2}), \tag{A.25}$$

where

$$\tau_0(u) = u^\top\{(\Omega^{[M]})^{-1} - A\}u,$$

$$\begin{aligned}\tau_1^\phi(u) = \tfrac{1}{3}\phi'''(1)L_2^{3[M]} + \left\{1 + \tfrac{2}{3}\phi'''(1)\right\} L(1)_2^3 \\ - \{1 + \phi'''(1)\}\left\{L(2)_{1,1}^{1,2} - 2L(3)_{1,1}^{1,1,1}\right\},\end{aligned}$$

$$\begin{aligned}\tau_2^\phi(u) = \tfrac{1}{12}\phi^{(4)}(1)L_3^{4[M]} - \left\{1 + \tfrac{4}{3}\phi'''(1) + \tfrac{1}{4}\phi^{(4)}(1)\right\} L(1)_3^4 \\ -\tfrac{1}{3}\left\{2\phi'''(1) + \phi^{(4)}(1)\right\} L(2)_{1,2}^{1,3} + \left\{1 + 2\phi'''(1) + \tfrac{1}{2}\phi^{(4)}(1)\right\} \\ \times \left\{L(2)_{2,1}^{2,2} + L(3)_{1,1}^{2,2,0} - 2L(3)_{1,2}^{1,2,1} - 2L(3)_{2,1}^{2,1,1}\right\} \\ + \left\{1 + 3\phi'''(1) + \phi^{(4)}(1)\right\}\{L(4) - 2L(5)\},\end{aligned}$$

$$\begin{aligned}A = \sum_{m=1}^{M} (I_{N^{[M]}-1} - 1_{N^{[M]}-1})\Big((1_{J_1}1_{J_1}^\top) \otimes \cdots \otimes (1_{J_{m-1}}1_{J_{m-1}}^\top) \otimes D_m^{-1} \\ \otimes (1_{J_{m+1}}1_{J_{m+1}}^\top) \otimes \cdots \otimes (1_{J_M}1_{J_M}^\top)\Big)(I_{N^{[M]}-1} - 1_{N^{[M]}-1})^\top,\end{aligned}$$

$$L_b^{a[M]} = \sum_{j_1=1}^{J_1} \cdots \sum_{j_M=1}^{J_M} \frac{u_{j_1\cdots j_M}^a}{Q(j_1, \ldots, j_M)^b},$$

$$L(1)_b^a = \sum_{m=1}^{M}\sum_{j_m=1}^{J_m} \frac{u_{\cdot(m,j_m)}^a}{p_{\cdot(m,j_m)}^b},$$

$$L(2)_{c,d}^{a,b} = \sum_{m=1}^{M}\sum_{j_1=1}^{J_1} \cdots \sum_{j_M=1}^{J_M} \frac{u_{\cdot(m,j_m)}^a u_{j_1\cdots j_M}^b}{p_{\cdot(m,j_m)}^c Q(j_1, \ldots, j_M)^d},$$

$$L(3)_{d,e}^{a,b,c} = \sum_{\ell=1}^{M-1}\sum_{m=\ell+1}^{M}\sum_{j_\ell=1}^{J_\ell}\sum_{j_m=1}^{J_m} \frac{u_{\cdot(\ell,j_\ell)}^a u_{\cdot(m,j_m)}^b u_{\cdot(\ell,j_\ell;m,j_m)}^c}{p_{\cdot(\ell,j_\ell)}^d p_{\cdot(m,j_m)}^e},$$

$$L(4) = \sum_{\ell=1}^{M-1} \sum_{m=\ell+1}^{M} \sum_{j_1=1}^{J_1} \cdots \sum_{j_M=1}^{J_M} \frac{u_{\cdot(\ell,j_\ell)} u_{\cdot(m,j_m)} u_{j_1\cdots j_M}^2}{p_{\cdot(\ell,j_\ell)} p_{\cdot(m,j_m)} Q(j_1,\ldots,j_M)},$$

$$L(5) = \begin{cases} \displaystyle\sum_{k=1}^{M-2} \sum_{\ell=k+1}^{M-1} \sum_{m=\ell+1}^{M} \sum_{j_k=1}^{J_k} \sum_{j_\ell=1}^{J_\ell} \sum_{j_m=1}^{J_m} \frac{u_{\cdot(k,j_k)} u_{\cdot(\ell,j_\ell)} u_{\cdot(m,j_m)} u_{\cdot(k,j_k;\ell,j_\ell;m,j_m)}}{p_{\cdot(k,j_k)} p_{\cdot(\ell,j_\ell)} p_{\cdot(m,j_m)}} & (M \geq 3) \\ 0 & (M = 2) \end{cases},$$

$1_{N^{[M]}-1}$ is an $(N^{[M]}-1)$-dimensional column vector with unit elements, $Q(j_1,\ldots,j_M)$ is defined by (4.12), $\Omega^{[M]}$ and D_m are defined in Theorem 4.3, and $\tau_3^{\phi}(u)$ is a homogeneous polynomial of degree 5 with respect to variables $u_{1\cdots1},\ldots,u_{J_1\cdots J_{M-1}J_M-1}$. By substituting the expanded expression of $\exp\{itC_{\phi}^{(1)[M]}(u)\}$ that is obtained by using (A.25) for $\exp\{itC_{\phi}^{(1)[M]}(u)\}$ in (A.24), $\psi_{\phi}(t)$ is expressed as follows.

$$\psi_{\phi}(t) = \left(|\Omega^{[M]}|/|\Lambda|\right)^{-1/2} \int_{-\infty}^{\infty} \cdots \int_{-\infty}^{\infty} h_0(u) g_{\phi}(u) du + O(n^{-2}), \tag{A.26}$$

where

$$h_0(u) = (2\pi)^{-(N^{[M]}-1)/2} |\Lambda|^{-1/2} \exp\left(-\frac{1}{2} u^{\top} \Lambda^{-1} u\right),$$

$$\begin{aligned} g_{\phi}(u) = 1 &+ n^{-1/2}\left\{h_1^{[M]}(u) + (it)\tau_1^{\phi}(u)\right\} \\ &+ n^{-1}\left\{h_2^{[M]}(u) + (it)\tau_1^{\phi}(u) h_1^{[M]}(u) + (it)\tau_2^{\phi}(u) + \tfrac{1}{2}(it)^2\left(\tau_1^{\phi}(u)\right)^2\right\} \\ &+ n^{-3/2} Q_0(u), \end{aligned}$$

where $\Lambda = (1-2it)^{-1}\left(\Omega^{[M]} - 2it\Omega^{[M]} A \Omega^{[M]}\right)$, and the degrees of all terms of polynomial $Q_0(u)$ are odd. Since $|\Omega^{[M]}|/|\Lambda| = (1-2it)^{f^{[M]}}$, by calculating the integral of (A.26), $\psi_{\phi}(t)$ is expanded as

$$\psi_{\phi}(t) = (1-2it)^{-f^{[M]}/2}\left\{1 + n^{-1}\sum_{\nu=0}^{3}(1-2it)^{-\nu} d_{\nu}^{\phi[M]} + O(n^{-2})\right\}, \tag{A.27}$$

where $d_{\nu}^{\phi[M]}$, $\nu = 0, 1, 2, 3$ are defined in Theorem 4.4. By inverting (A.27), we obtain (4.20). We have completed the proof of Theorem 4.4.

A.5 Proof of Theorem 5.2

By the transformation (5.7), the statistic C_ϕ can be rewritten as

$$C_\phi(U) = 2n \sum_{j=1}^{J} \sum_{k=1}^{K} \sum_{\ell=1}^{L} q_{jk\ell} G_{jk\ell} \phi(D_{jk\ell}),$$

where

$$G_{jk\ell} = \left(1 + \frac{U_{j\cdot\ell}}{\sqrt{n} p_{j\cdot\ell}}\right)\left(1 + \frac{U_{\cdot k\ell}}{\sqrt{n} p_{\cdot k\ell}}\right)\left(1 + \frac{U_{\cdot\cdot\ell}}{\sqrt{n} p_{\cdot\cdot\ell}}\right)^{-1},$$

$$j = 1, \ldots, J, k = 1, \ldots, K, \ell = 1, \ldots, L,$$

$$D_{jk\ell} = \left(1 + \frac{U_{jk\ell}}{\sqrt{n} q_{jk\ell}}\right) G_{jk\ell}^{-1}, \quad j = 1, \ldots, J, k = 1, \ldots, K, \ell = 1, \ldots, L,$$

$U_{j\cdot\ell} = \sum_{k=1}^{K} U_{jk\ell}$, $U_{\cdot k\ell} = \sum_{j=1}^{J} U_{jk\ell}$, and $U_{\cdot\cdot\ell} = \sum_{j=1}^{J} \sum_{k=1}^{K} U_{jk\ell}$.

Since J_1^* is multivariate Edgeworth expansion, J_1^* is represented as

$$J_1^* = \int \cdots \int_{B_\phi(x)} f(u) \left\{1 + n^{-1/2} h_1(u) + n^{-1} h_2(u) + n^{-3/2} h_3(u)\right\} du + O(n^{-2}),$$

where

$$B_\phi(x) = \{u : C_\phi(u) \le x | H_0\},$$

$du = du_{111} \cdots du_{JK,L-1}$, and f, h_1, h_2 and h_3 are defined by (5.10), (5.12), (5.13) and (5.14), respectively.

We derive an approximation of the characteristic function of $C_\phi(U)$, that is,

$$\begin{aligned} \psi_\phi(t) = \int_{\infty}^{\infty} \cdots \int_{-\infty}^{\infty} & \exp\{it C_\phi(u)\} f(u) \\ & \times \left\{1 + n^{-1/2} h_1(u) + n^{-1} h_2(u) + n^{-3/2} h_3(u)\right\} du \end{aligned} \quad \text{(A.28)}$$

up to order $n^{-3/2}$. By the assumption of $\phi(t)$, we obtain

$$C_\phi(u) = \tau_0(u) + n^{-1/2} \tau_1^\phi(u) + n^{-1} \tau_2^\phi(u) + n^{-3/2} \tau_3^\phi(u) + O(n^{-2}), \quad \text{(A.29)}$$

where

$$\tau_0(u) = u' \Omega^{-1} u + M(1)_2 - M(2)_{0,2},$$

$$\tau_1^{\phi}(u) = \frac{1}{3}\phi'''(1)M_2^3 + \left\{\frac{4}{3}\phi'''(1) + 1\right\}M(1)_3 + \left\{\frac{2}{3}\phi'''(1) + 1\right\}M(2)_{0,3}$$
$$+ \{\phi'''(1) + 1\}\{-2M(2)_{1,2} + 2M(3)_{0,1,1} + M(3)_{1,0,2} - M(4)_{0,0,1,2}\},$$

$$\tau_2^{\phi}(u) = \frac{1}{12}\phi^{(4)}(1)M_3^4 + \left\{\frac{3}{4}\phi^{(4)}(1) + \frac{4}{3}\phi'''(1)\right\}M(1)_4$$
$$- \left\{\frac{1}{4}\phi^{(4)}(1) + \frac{4}{3}\phi'''(1) + 1\right\}M(2)_{0,4}$$
$$- \left\{\frac{3}{2}\phi^{(4)}(1) + 4\phi'''(1) + 1\right\}M(2)_{2,2}$$
$$+ \left\{\frac{1}{2}\phi^{(4)}(1) + 2\phi'''(1) + 1\right\}$$
$$\times \left\{2M(2)_{1,3} + M(3)_{0,2,0} + M(4)_{0,0,2,2} - 2M(4)_{0,1,2,1}\right\}$$
$$+ \left\{\phi^{(4)}(1) + 2\phi'''(1)\right\}\left\{\frac{1}{2}M(3)_{2,0,2} + \frac{1}{3}M(3)_{1,0,3} - \frac{1}{3}M(4)_{0,0,1,3}\right\}$$
$$+ \left\{\phi^{(4)}(1) + 3\phi'''(1) + 1\right\}\left\{M(3)_{0,1,2} + 2M(3)_{1,1,1} - M(4)_{1,0,1,2}\right\},$$

$$M(1)_a = \sum_{\ell=1}^{L} \frac{u_{\cdot\cdot\ell}^a}{p_{\cdot\cdot\ell}^{a-1}},$$

$$M(2)_{a,b} = \sum_{j=1}^{J}\sum_{\ell=1}^{L}\left(\frac{u_{\cdot\cdot\ell}}{p_{\cdot\cdot\ell}}\right)^a \frac{u_{j\cdot\ell}^b}{p_{j\cdot\ell}^{b-1}} + \sum_{k=1}^{K}\sum_{\ell=1}^{L}\left(\frac{u_{\cdot\cdot\ell}}{p_{\cdot\cdot\ell}}\right)^a \frac{u_{\cdot k\ell}^b}{p_{\cdot k\ell}^{b-1}},$$

$$M(3)_{a,b,c} = \sum_{j=1}^{J}\sum_{k=1}^{K}\sum_{\ell=1}^{L}\left(\frac{u_{\cdot\cdot\ell}}{p_{\cdot\cdot\ell}}\right)^a \left(\frac{u_{j\cdot\ell}u_{\cdot k\ell}}{p_{j\cdot\ell}p_{\cdot k\ell}}\right)^b \frac{u_{jk\ell}^c}{q_{jk\ell}^{c-1}},$$

$$M(4)_{a,b,c,d} = \sum_{j=1}^{J}\sum_{k=1}^{K}\sum_{\ell=1}^{L}\left(\frac{u_{\cdot\cdot\ell}}{p_{\cdot\cdot\ell}}\right)^a$$
$$\times \left\{\left(\frac{u_{j\cdot\ell}}{p_{j\cdot\ell}}\right)^b \left(\frac{u_{\cdot k\ell}}{p_{\cdot k\ell}}\right)^c + \left(\frac{u_{j\cdot\ell}}{p_{j\cdot\ell}}\right)^c \left(\frac{u_{\cdot k\ell}}{p_{\cdot k\ell}}\right)^b\right\} \frac{u_{jk\ell}^d}{q_{jk\ell}^{d-1}},$$

where Ω is defined by (5.11) and $\tau_3^{\phi}(u)$ is a homogeneous polynomial of degree 5 with respect to variables $u_{111}, \ldots, u_{JK,L-1}$. By substituting the expanded expression of $\exp\{itC_{\phi}(u)\}$ that is obtained by using (A.29) for $\exp\{itC_{\phi}(u)\}$ in (A.28), $\psi_{\phi}(t)$ is expressed as follows.

$$\psi_{\phi}(t) = \left(\frac{|\Omega|}{|\Lambda|}\right)^{-1/2} \int_{-\infty}^{\infty} \cdots \int_{-\infty}^{\infty} h_0(u) g_{\phi}(u) du + O(n^{-2}), \tag{A.30}$$

where

$$h_0(u) = (2\pi)^{-(N-1)/2}|\Lambda|^{-1/2}\exp\left(-\frac{1}{2}u^\top\Lambda^{-1}u\right),$$

$$\begin{aligned} g_\phi(u) &= 1 + n^{-1/2}\{h_1(u) + (it)\tau_1^\phi(u)\} \\ &\quad + n^{-1}\{h_2(u) + (it)\tau_1^\phi(u)h_1(u) + (it)\tau_2\phi(u) + \tfrac{1}{2}(it)^2(\tau_1^\phi(u))^2\} \\ &\quad + n^{-3/2}Q_0(u), \end{aligned}$$

$$\Lambda = (1-2it)^{-1}(\Omega - 2it\Omega\Xi\Omega),$$

$$\begin{aligned} \Xi &= (I_{N-1} \ - 1_{N-1})\left((1_K1_K^\top)\otimes D_1^{-1}\oplus\cdots\oplus(1_K1_K^\top)\otimes D_J^{-1}\right. \\ &\quad \left. + (1_J1_J^\top)\otimes D_{(2)}^{-1} - (1_J1_J^\top)\otimes(1_K1_K^\top)\otimes D_{(3)}^{-1}\right)(I_{N-1} \ - 1_{N-1})^\top, \end{aligned}$$

$$D_j = \mathrm{diag}(p_{j\cdot1},\ldots,p_{j\cdot L})\ (j=1,\ldots,J),$$

$$D_{(2)} = \mathrm{diag}(p_{\cdot11},\ldots,p_{\cdot1L},\ldots,p_{\cdot K1},\ldots,p_{\cdot KL}),$$

$$D_{(3)} = \mathrm{diag}(p_{\cdot\cdot1},\ldots,p_{\cdot\cdot L}),$$

where N is given by (5.9), $\otimes$ and $\oplus$ denote the Kronecker product and direct sum of matrices, and the degrees of all terms of polynomial $Q_0(u)$ are odd.

We calculate the integral of (A.30) by using the equation $|\Omega|/|\Lambda| = (1-2it)^M$. Then, $\psi_\phi(t)$ are expanded as follows:

$$\psi_\phi(t) = (1-2it)^{-M/2}\left\{1 + n^{-1}\sum_{j=0}^{3}(1-2it)^{-j}v_j^\phi + O(n^{-2})\right\}, \qquad \text{(A.31)}$$

where v_j^ϕ $(j=0,1,2,3)$ are defined in (5.15). Results of Theorem 5.2 are derived by inverting (A.31).

A.6 Proof of Theorem 6.2

By transformation (6.10), statistic C_ϕ^g can be rewitten as

$$C_\phi^g(W) = 2\sum_{\alpha=1}^{N} n_\alpha \left\{ \hat{\pi}_\alpha^g(W)\phi\left(\frac{\pi_\alpha^g + W_\alpha(\sqrt{n_\alpha})^{-1}}{\hat{\pi}_\alpha^g(W)}\right) + \left(1-\hat{\pi}_\alpha^g(W)\right)\phi\left(\frac{1-\pi_\alpha^g - W_\alpha(\sqrt{n_\alpha})^{-1}}{1-\hat{\pi}_\alpha^g(W)}\right)\right\}.$$

If we regard

$$h^g(w)\left\{1+n^{-1/2}h_1^g(w)+n^{-1}h_2^g(w)+n^{-3/2}h_3^g(w)\right\}$$

as the continuous density function of W, then we can regard

$$J_1^{g,\phi}(x) = \int\cdots\int_{U_\phi^g(x)} h^g(w)\left\{1+n^{-1/2}h_1^g(w)+n^{-1}h_2^g(w)+n^{-3/2}h_3^g(w)\right\}dw$$

as the distribution function of $C_\phi^g(W)$, where

$$U_\phi^g(x) = \{w=(w_1,\ldots,w_N)^\top : C_\phi^g(w)\le x\}.$$

So, the characteristic function of $C_\phi^g(W)$ is calculated as

$$\psi_\phi^g(w) = \int_{-\infty}^{\infty}\cdots\int_{-\infty}^{\infty}\left[\exp\{iuC_\phi^g(w)\}\right]h^g(w)\left\{1+n^{-1/2}h_1^g(w)+n^{-1}h_2^g(w) + n^{-3/2}h_3^g(w)\right\}dw. \tag{A.32}$$

We can expand $C_\phi^g(w)$ as

$$C_\phi^g(w) = \tau_0^g(w) + n^{-1/2}\tau_1^{g,\phi}(w) + n^{-1}\tau_2^{g,\phi}(w) + n^{-3/2}\tau_3^{g,\phi}(w) + O(n^{-2}), \tag{A.33}$$

where

$$\tau_0^g(w) = w^\top(\Omega^{-1}-\Xi)w,$$

$\Xi = (\xi_{\alpha\beta})$ is a $N\times N$ matrix,

$$\xi_{\alpha\beta} = \frac{\sqrt{\mu_\alpha}G_1(\alpha)}{\pi_\alpha^g(1-\pi_\alpha^g)}\frac{\sqrt{\mu_\beta}G_1(\beta)}{\pi_\beta^g(1-\pi_\beta^g)}\sigma_{\alpha\beta},\quad \alpha,\beta=1,\ldots,N,$$

$$\tau_1^{g,\phi}(w) = \sum_{a=0}^{3}\left(\sum_{\alpha=1}^{N} B_{a+1}^1(\alpha)C_{1(\alpha)}(w)^{3-a}w_\alpha^a\right),$$

$$
\begin{aligned}
\tau_2^{g,\phi}(w) = &\sum_{a=0}^{2}\left(\sum_{\alpha=1}^{N} B_{a+1}^2(\alpha) C_{2(\alpha)}(w)^{2-a} C_{1(\alpha)}(w)^{2a}\right) \\
&+\sum_{a=0}^{1}\left(\sum_{\alpha=1}^{N} B_{a+4}^2(\alpha) C_{2(\alpha)}(w)^{1-a} C_{1(\alpha)}(w)^{1+2a} w_\alpha\right) \\
&+\sum_{a=0}^{1}\left(\sum_{\alpha=1}^{N} B_{a+6}^2(\alpha) C_{2(\alpha)}(w)^{1-a} C_{1(\alpha)}(w)^{2a} w_\alpha^2\right) \\
&+\sum_{\alpha=1}^{N} B_8^2(\alpha) C_{1(\alpha)}(w) w_\alpha^3 + \sum_{\alpha=1}^{N} B_9^2(\alpha) w_\alpha^4,
\end{aligned}
$$

$$
\begin{aligned}
B_1^1(\alpha) = \frac{\mu_\alpha}{3(\pi_\alpha^g)^2(1-\pi_\alpha^g)^2}\left\{3\pi_\alpha^g(1-\pi_\alpha^g)G_1(\alpha)G_2(\alpha)\right. \\
\left. -(3+\phi'''(1))(1-2\pi_\alpha^g)G_1(\alpha)^3\right\},
\end{aligned}
$$

$$
B_2^1(\alpha) = \frac{\sqrt{\mu_\alpha}}{(\pi_\alpha^g)^2(1-\pi_\alpha^g)^2}\left\{-\pi_\alpha^g(1-\pi_\alpha^g)G_2(\alpha) + (2+\phi'''(1))(1-2\pi_\alpha^g)G_1(\alpha)^2\right\},
$$

$$
B_3^1(\alpha) = -\frac{\left(1+\phi'''(1)\right)(1-2\pi_\alpha^g)G_1(\alpha)}{(\pi_\alpha^g)^2(1-\pi_\alpha^g)^2}, \quad B_4^1(\alpha) = \frac{\phi'''(1)(1-2\pi_\alpha^g)}{3\sqrt{\mu_\alpha}(\pi_\alpha^g)^2(1-\pi_\alpha^g)^2},
$$

$$
B_1^2(\alpha) = \frac{\mu_\alpha G_1(\alpha)^2}{\pi_\alpha^g(1-\pi_\alpha^g)}, \quad B_2^2(\alpha) = 3B_1^1(\alpha),
$$

$$
\begin{aligned}
B_3^2(\alpha) = &\frac{\mu_\alpha}{12(\pi_\alpha^g)^3(1-\pi_\alpha^g)^3}\{(\pi_\alpha^g)^2(1-\pi_\alpha^g)^2(3G_2(\alpha)^2 + 4G_1(\alpha)G_3(\alpha)) \\
&-6(3+\phi'''(1))\pi_\alpha^g(1-\pi_\alpha^g)(1-2\pi_\alpha^g)G_1(\alpha)^2G_2(\alpha) \\
&+(12+8\phi'''(1)+\phi^{(4)}(1))(1-3\pi_\alpha^g+3(\pi_\alpha^g)^2)G_1(\alpha)^4\},
\end{aligned}
$$

$$
B_4^2(\alpha) = 2B_2^1(\alpha),
$$

$$
\begin{aligned}
B_5^2(\alpha) = &\frac{\sqrt{\mu_\alpha}}{3(\pi_\alpha^g)^3(1-\pi_\alpha^g)^3}\{-(\pi_\alpha^g)^2(1-\pi_\alpha^g)^2G_3(\alpha) \\
&+3(2+\phi'''(1))\pi_\alpha^g(1-\pi_\alpha^g)(1-2\pi_\alpha^g)G_1(\alpha)G_2(\alpha) \\
&-(6+6\phi'''(1)+\phi^{(4)}(1))(1-3\pi_\alpha^g+3(\pi_\alpha^g)^2)G_1(\alpha)^3\},
\end{aligned}
$$

$$
B_6^2(\alpha) = B_3^1(\alpha),
$$

$$B_7^2(\alpha) = \frac{1}{2(\pi_\alpha^g)^3(1-\pi_\alpha^g)^3}\{-(1+\phi'''(1))\pi_\alpha^g(1-\pi_\alpha^g)(1-2\pi_\alpha^g)G_2(\alpha)$$
$$+(2+4\phi'''(1)+\phi^{(4)}(1))(1-3\pi_\alpha^g+3(\pi_\alpha^g)^2)G_1(\alpha)^2\},$$

$$B_8^2(\alpha) = -\frac{(2\phi'''(1)+\phi^{(4)}(1))(1-3\pi_\alpha^g+3(\pi_\alpha^g)^2)G_1(\alpha)}{3\sqrt{\mu_\alpha}(\pi_\alpha^g)^3(1-\pi_\alpha^g)^3},$$

$$B_9^2(\alpha) = \frac{\phi^{(4)}(1)(1-3\pi_\alpha^g+3(\pi_\alpha^g)^2)}{12\mu_\alpha(\pi_\alpha^g)^3(1-\pi_\alpha^g)^3},$$

$$C_{1(\alpha)}(w) = \sum_{m=1}^{p} x_{\alpha m}\left(\sum_{k=1}^{p}\kappa^{m,k}M_k(w)\right), \ \alpha = 1,\ldots,N,$$

$$C_{2(\alpha)}(w) = \sum_{m=1}^{p} x_{\alpha m}\left\{\sum_{k=1}^{p} M^{m,k}(w)M_k(w)\right.$$
$$\left.+\frac{1}{2}\sum_{k_1=1}^{p}\cdots\sum_{k_5=1}^{p}\kappa^{m,k_3}\kappa^{k_1,k_4}\kappa^{k_2,k_5}\kappa_{k_3,k_4,k_5}M_{k_1}(w)M_{k_2}(w)\right\},$$
$$\alpha = 1,\ldots,N,$$

$$M_k(w) = \sum_{\lambda=1}^{N}\sqrt{\mu_\lambda}x_{\lambda k}G_1(\lambda)\{\pi_\lambda^g(1-\pi_\lambda^g)\}^{-1}w_\lambda, \quad k = 1,\ldots,p,$$

$$J_{i,j}(w) = \sum_{\lambda=1}^{N}\sqrt{\mu_\lambda}x_{\lambda i}x_{\lambda j}\left\{-\frac{(1-2\pi_\lambda^g)G_1(\lambda)^2}{(\pi_\lambda^g)^2(1-\pi_\lambda^g)^2}+\frac{G_2(\lambda)}{\pi_\lambda^g(1-\pi_\lambda^g)}\right\}w_\lambda,$$
$$i,j = 1,\ldots,p,$$

$$\kappa_{i,j,k} = \sum_{\lambda=1}^{N}\mu_\lambda x_{\lambda i}x_{\lambda j}x_{\lambda k}\left\{2\frac{(1-2\pi_\lambda^g)G_1(\lambda)^3}{(\pi_\lambda^g)^2(1-\pi_\lambda^g)^2}-3\frac{G_1(\lambda)G_2(\lambda)}{\pi_\lambda^g(1-\pi_\lambda^g)}\right\},$$
$$i,j,k = 1,\ldots,p,$$

$$G_i(\alpha) = u^{(i)}(x_\alpha^\top\beta), \qquad \alpha = 1,\cdots,N, \ i = 1,2,3,$$

$J(w) = (J_{i,j}(w))$ is a $p \times p$ matrix, $M^{i,j}(w)$ is the (i,j)-element of matrix $K^{-1}J(w)K^{-1}$, Ω is defined by (6.12), $\sigma_{\alpha\beta}$ and $K^{-1} = (\kappa^{i,j})$ are defined in Theorem 6.2, and $\tau_3^{g,\phi}(w)$ is a homogeneous polynomial of degree 5 with respect to variables $w_1,\ldots,w_N$. Then, from (6.11), (A.32) and (A.33), we obtain

$$\psi_\phi^g(w) = (1-2iw)^{-(N-p)/2}\int_{-\infty}^{\infty}\cdots\int_{-\infty}^{\infty}(2\pi)^{-N/2}|\Lambda|^{-1/2}\left\{\exp\left(\frac{1}{2}w^\top\Lambda^{-1}w\right)\right\}$$
$$\times\{1+n^{-1/2}D_1(w)+n^{-1}D_2(w)+n^{-3/2}D_3(w)\}dw+O(n^{-2}), \quad (A.34)$$

where $\Lambda = (1-2iw)^{-1}(\Omega - 2iw\Omega\Xi\Omega)$, $D_1(w) = h_1^g(w) + (iw)\tau_1^{g,\phi}(w)$,

$$D_2(w) = h_2^g(w) + (iw)\tau_1^{g,\phi}(w)h_1^g(w) + (iw)\tau_2^{g,\phi}(w) + \frac{1}{2}(iw)^2\left\{\tau_1^{g,\phi}(w)\right\}^2,$$

and degrees of all terms of polynomial $D_3(w)$ are odd. Therefore, by carrying out the integration of (A.34), the characteristic function $\psi_\phi^g(w)$ is expanded as

$$\psi_\phi^g(w) = (1-2iw)^{-(N-p)/2}\left[1+n^{-1}\sum_{j=0}^{3}(1-2iw)^{-j}v_j^{g,\phi}+O(n^{-2})\right]. \quad (A.35)$$

By inverting (A.35), we obtain (6.13). We have completed the proof of Theorem 6.2.

A.7 Proof of Theorem 7.1

If we set $\tilde{Y} = (Y_1,\ldots,Y_{N-1})^\top$, the characteristic function $c(s)$ of $\tilde{Y}$ is given by

$$\begin{aligned} c(s) &= \sum_{\tilde{y}\in S_0}\exp(is^\top\tilde{y})\Pr\{\tilde{Y}=\tilde{y}\mid H_0^L\} \\ &= \left\{\sum_{\alpha=1}^{N-1}\pi_\alpha(\theta)\exp(is_\alpha)+\pi_N(\theta)\right\}^n, \end{aligned}$$

where $s=(s_1,\ldots,s_{N-1})^\top$. For each $u = n^{-1/2}(\tilde{y}-n\tilde{\pi}(\theta))\in S_1$, we evaluate $\Pr\{U=u\mid H_0^L\}$ in the following way:

$$\begin{aligned} \Pr\{U=u\mid H_0^L\} &= \Pr\{\tilde{Y}=\tilde{y}\mid H_0^L\} \\ &= (2\pi)^{-(N-1)}\int_{-\pi}^{\pi}\cdots\int_{-\pi}^{\pi}c(s)\exp(-is^\top\tilde{y})ds \\ &= (2\pi n^{1/2})^{-(N-1)}Q, \end{aligned}$$

where

$$Q = \int_{-n^{1/2}\pi}^{n^{1/2}\pi}\cdots\int_{-n^{1/2}\pi}^{n^{1/2}\pi}q(s)\exp(-is^\top u)ds \quad (A.36)$$

and

$$q(s) = c(n^{-1/2}s)\exp\left(-in^{1/2}s^\top\tilde{\pi}(\theta)\right).$$

We can write

$$q(s) = \left\{\exp\left(-\frac{1}{2}s^\top \Omega s\right)\right\}\left\{1 + \sum_{\upsilon=1}^{3} n^{-\upsilon/2} b_\upsilon(s) + O(n^{-2})\right\} \tag{A.37}$$

for large n and fixed s, where

$$b_1(s) = \frac{i^3}{6}\left\{\sum_{\alpha=1}^{N-1} \pi_\alpha(\theta)s_\alpha^3 - 3(s^\top \tilde{\pi}(\theta))s^\top \Omega s - (s^\top \tilde{\pi}(\theta))^3\right\},$$

$$\begin{aligned} b_2(s) = \frac{1}{2}\{b_1(s)\}^2 + \frac{i^4}{24}\Bigg\{&\sum_{\alpha=1}^{N-1}\pi_\alpha(\theta)s_\alpha^4 - 4(s^\top \tilde{\pi}(\theta))\sum_{\alpha=1}^{N-1}\pi_\alpha(\theta)s_\alpha^3 \\ &-3(s^\top \Omega s)^2 + 6(s^\top \tilde{\pi}(\theta))^2 s^\top \Omega s + 3(s^\top \tilde{\pi}(\theta))^4\Bigg\} \end{aligned}$$

and

$$\begin{aligned} b_3(s) = -\frac{1}{3}\{b_1(s)\}^3 + b_1(s)b_2(s) + \frac{i^5}{120}\Bigg\{&\sum_{\alpha=1}^{N-1}\pi_\alpha(\theta)s_\alpha^5 - 5(s^\top \tilde{\pi}(\theta))\sum_{\alpha=1}^{N-1}\pi_\alpha(\theta)s_\alpha^4 \\ &-10(s^\top \Omega s)\sum_{\alpha=1}^{N-1}\pi_\alpha(\theta)s_\alpha^3 + 10(s^\top \tilde{\pi}(\theta))^2\sum_{\alpha=1}^{N-1}\pi_\alpha(\theta)s_\alpha^3 \\ &+30(s^\top \tilde{\pi}(\theta))(s^\top \Omega s)^2 - 6(s^\top \tilde{\pi}(\theta))^5\Bigg\}. \end{aligned}$$

By using (A.37), Q given by (A.36) is divided into three parts.

$$Q = Q_1 + Q_2 - Q_3,$$

where

$$Q_1 = \int_{-\infty}^{\infty}\cdots\int_{-\infty}^{\infty}\{\exp(-is^\top u)\}\left\{\exp\left(-\frac{1}{2}s^\top \Omega s\right)\right\}\left\{1 + \sum_{\upsilon=1}^{3} n^{-\upsilon/2} b_\upsilon(s)\right\} ds,$$

$$Q_2 = \int_{-\infty}^{\infty}\cdots\int_{-\infty}^{\infty}\{\exp(-is^\top u)\}\left\{\exp\left(-\frac{1}{2}s^\top \Omega s\right)\right\} O(n^{-2}) ds,$$

$$\begin{aligned} Q_3 = \int\cdots\int_{F^c}&\{\exp(-is^\top u)\}\left\{\exp\left(-\frac{1}{2}s^\top \Omega s\right)\right\} \\ &\times\left\{1 + \sum_{\upsilon=1}^{3} n^{-\upsilon/2} b_\upsilon(s) + O(n^{-2})\right\} ds, \end{aligned}$$

and

$$F = [-n^{1/2}\pi, n^{1/2}\pi] \times \cdots \times [-n^{1/2}\pi, n^{1/2}\pi].$$

Evaluation $Q = Q_1 + O(n^{-2})$ is derived by using $Q_2 = O(n^{-2})$ and $Q_3 = o(n^{-2})$. Therefore, we obtain the following evaluation:

$$\Pr\{U = u | H_0^L\} = (2\pi n^{1/2})^{-(N-1)} \left[\int_{-\infty}^{\infty} \cdots \int_{-\infty}^{\infty} \{\exp(-is^\top u)\} \right.$$
$$\left. \times \left\{ \exp\left(-\frac{1}{2}s^\top \Omega s\right)\right\} \left\{1 + \sum_{v=1}^{3} n^{-v/2} b_v(s)\right\} ds + O(n^{-2}) \right].$$

By calculating the integral of the above expression, we obtain the results of Theorem 7.1.

A.8 Proof of Theorem 7.2

We put $\hat{\pi}_\alpha = \pi_\alpha(\hat{\theta})$, $\alpha = 1, \ldots, N$, where $\hat{\theta}$ is MLE of θ under H_0^L. Since $\hat{\theta}$ is a function of Y, $\hat{\pi}_\alpha$ is also a function of Y. Then, $\hat{\pi}_\alpha$ is a function of U by the transformation (7.5), and we put it $\hat{\pi}_\alpha(U)$. Therefore, G^2 given by (7.2) is rewritten as

$$G^2(U) = 2n \sum_{\alpha=1}^{N} (\pi_\alpha(\theta) + n^{-1/2} U_\alpha) \ln\left(\frac{\pi_\alpha(\theta) + n^{-1/2} U_\alpha}{\hat{\pi}_\alpha(U)}\right).$$

If we regard U as a continuous random variable and let

$$f(u)\left\{1 + n^{-1/2} h_1(u) + n^{-1} h_2(u) + n^{-3/2} h_3(u)\right\}$$

be a probability density function of U. Then, we can regard

$$W_1 = \int \cdots \int_{B(x)} f(u)\left\{1 + n^{-1/2} h_1(u) + n^{-1} h_2(u) + n^{-3/2} h_3(u)\right\} du$$

as a distribution function of G^2, where $du = du_1 \cdots du_{N-1}$ and $B(x) = \{u = (u_1, \ldots, u_{N-1})^\top : G^2(u) \le x\}$. So, the characteristic function of $G^2(U)$ is calculated as

$$\psi(t) = \int_{-\infty}^{\infty} \cdots \int_{-\infty}^{\infty} \exp\{itG^2(u)\} f(u) \left\{1 + n^{-1/2} h_1(u) + n^{-1} h_2(u) \right.$$
$$\left. + n^{-3/2} h_3(u)\right\} du. \tag{A.38}$$

We can expand $G^2(u)$ as

$$G^2(u) = \tau_0(u) + n^{-1/2}\tau_1(u) + n^{-1}\tau_2(u) + n^{-3/2}\tau_3(u) + O(n^{-2}), \tag{A.39}$$

where

$$\tau_0(u) = u^\top(\Omega^{-1} - \Xi)u,$$

$$\begin{aligned}\tau_1(u) = -\frac{1}{3}\Bigg\{&\sum_{\alpha=1}^{N} u_\alpha^3\pi_\alpha^{-2} - \sum_{\alpha=1}^{N}\sum_{\beta=1}^{N}\sum_{\gamma=1}^{N} u_\alpha u_\beta u_\gamma \sum_{\lambda=1}^{N} \pi_\lambda\sigma_{\alpha\lambda}\sigma_{\beta\lambda}\sigma_{\gamma\lambda}\\
&+ 3\sum_{\alpha=1}^{N}\sum_{\beta=1}^{N} u_\alpha u_\beta \sum_{\lambda=1}^{N}\pi_\lambda\sigma_{\alpha\lambda}\sigma_{\beta\lambda} \times \sum_{\gamma=1}^{N} u_\gamma \sum_{\mu=1}^{N}\pi_\mu\sigma_{\gamma\mu}\\
&- 2\left(\sum_{\alpha=1}^{N} u_\alpha \sum_{\lambda=1}^{N}\pi_\lambda\sigma_{\alpha\lambda}\right)^3\Bigg\},\end{aligned}$$

$$\begin{aligned}\tau_2(u) = &\frac{1}{6}\sum_{\alpha=1}^{N} u_\alpha^4\pi_\alpha^{-3} + \frac{1}{12}\sum_{\alpha=1}^{N}\sum_{\beta=1}^{N}\sum_{\gamma=1}^{N}\sum_{\varepsilon=1}^{N} u_\alpha u_\beta u_\gamma u_\varepsilon \sum_{\lambda=1}^{N}\pi_\lambda\sigma_{\alpha\lambda}\sigma_{\beta\lambda}\sigma_{\gamma\lambda}\sigma_{\varepsilon\lambda}\\
&-\frac{1}{4}\sum_{\alpha=1}^{N}\sum_{\beta=1}^{N}\sum_{\gamma=1}^{N}\sum_{\varepsilon=1}^{N} u_\alpha u_\beta u_\gamma u_\varepsilon \sum_{\lambda=1}^{N}\sum_{\mu=1}^{N}\pi_\lambda\pi_\mu\sigma_{\lambda\mu}\sigma_{\alpha\lambda}\sigma_{\beta\lambda}\sigma_{\gamma\mu}\sigma_{\varepsilon\mu}\\
&+\frac{2}{3}\sum_{\alpha=1}^{N}\sum_{\beta=1}^{N}\sum_{\gamma=1}^{N} u_\alpha u_\beta u_\gamma \sum_{\lambda=1}^{N}\pi_\lambda\sigma_{\alpha\lambda}\sigma_{\beta\lambda}\sigma_{\gamma\lambda} \times \sum_{\varepsilon=1}^{N} u_\varepsilon \sum_{\mu=1}^{N}\pi_\mu\sigma_{\varepsilon\mu}\\
&-\frac{1}{4}(1 + A_3)\left(\sum_{\alpha=1}^{N}\sum_{\beta=1}^{N} u_\alpha u_\beta \sum_{\lambda=1}^{N}\pi_\lambda\sigma_{\alpha\lambda}\sigma_{\beta\lambda}\right)^2\\
&+\frac{1}{2}\sum_{\alpha=1}^{N}\sum_{\beta=1}^{N} u_\alpha u_\beta \sum_{\lambda=1}^{N}\pi_\lambda\sigma_{\alpha\lambda}\sigma_{\beta\lambda} \times \sum_{\gamma=1}^{N}\sum_{\varepsilon=1}^{N} u_\gamma u_\varepsilon \sum_{\mu=1}^{N}\sum_{\nu=1}^{N}\pi_\mu\pi_\nu\sigma_{\mu\nu}\sigma_{\gamma\mu}\sigma_{\varepsilon\mu}\\
&-\sum_{\alpha=1}^{N}\sum_{\beta=1}^{N} u_\alpha u_\beta\sigma_{\alpha\beta} \times \left(\sum_{\gamma=1}^{N} u_\gamma \sum_{\lambda=1}^{N}\pi_\lambda\sigma_{\gamma\lambda}\right)^2 - \frac{1}{2}\left(\sum_{\alpha=1}^{N} u_\alpha \sum_{\lambda=1}^{N}\pi_\lambda\sigma_{\alpha\lambda}\right)^4,\end{aligned}$$

Ξ is an $(N-1)\times(N-1)$ symmetric matrix whose (α, β)-element is $\xi_{\alpha\beta} = \sigma_{\alpha\beta} - \sigma_{\alpha N} - \sigma_{N\beta} + \sigma_{NN}$, $\alpha = 1, \ldots, N-1$, $\beta = 1, \ldots, N-1$, Ω is given by (7.6), A_3 is defined in Theorem 7.2, and $\tau_3(u)$ is a homogeneous polynomial of degree 5 with respect to the variables $u_1, \ldots, u_{N-1}$.

By replacing $G^2(u)$ of (A.38) with (A.39), we obtain

$$\psi(t) = \left(\frac{|\,\Omega\,|}{|\,\Lambda\,|}\right)^{-1/2}\int_{-\infty}^{\infty}\cdots\int_{-\infty}^{\infty} h_0(u)g(u)du + O(n^{-2}), \tag{A.40}$$

where

$$h_0(u) = (2\pi)^{-(N-1)/2}|\Lambda|^{-1/2}\exp\left(-\frac{1}{2}u^\top\Lambda^{-1}u\right),$$

$$\begin{aligned} g(u) = 1 &+ n^{-1/2}\{h_1(u) + (it)\tau_1(u)\} \\ &+ n^{-1}\{h_2(u) + (it)\tau_1(u)h_1(u) + (it)\tau_2(u) + \tfrac{1}{2}(it)^2(\tau_1(u))^2\} \\ &+ n^{-3/2}Q_0(u), \end{aligned}$$

$$\Lambda = (1-2it)^{-1}(\Omega - 2it\Omega\Xi\Omega),$$

and the degrees of all terms of polynomial $Q_0(u)$ are odd. We compute the integral in (A.40). Then we obtain the following expanded expression (A.41) since $|\Omega| / |\Lambda| = (1-2it)^{N-p-1}$.

$$\psi(t) = (1-2it)^{-(N-p-1)/2}\left\{1 + n^{-1}\sum_{j=0}^{1}(1-2it)^{-j}d_j + O(n^{-2})\right\}. \quad \text{(A.41)}$$

By inverting (A.41), the results of Theorem 7.2 are derived.

A.9 Proof of Theorems 8.1 and 8.2

If we assume that statistic C_ϕ has a continuous distribution, the characteristic function of C_ϕ is evaluated as

$$\psi_\phi^C(t) = (1-2it)^{-\lambda/2} + \frac{1}{n}\sum_{j=0}^{3}(1-2it)^{-(\lambda+2j)/2}r_j^\phi + o(n^{-1}),$$

where

$$\begin{aligned} r_0^\phi &= \frac{1}{12}\Big[-(S-1)\Big], \\ r_1^\phi &= \frac{1}{24}\Big[3(3S-k^2-2k) + 6\phi'''(1)(S-k^2) \\ &\quad + \{\phi'''(1)\}^2(5S-3k^2-6k+4) - 3\phi^{(4)}(1)(S-2k+1)\Big], \\ r_2^\phi &= \frac{1}{24}\Big[-6(2S-k^2-2k+1) - 4\phi'''(1)(4S-3k^2-3k+2) \\ &\quad -2\{\phi'''(1)\}^2(5S-3k^2-6k+4) + 3\phi^{(4)}(1)(S-2k+1)\Big], \\ r_3^\phi &= \frac{1}{24}\Big[\{1+\phi'''(1)\}^2(5S-3k^2-6k+4)\Big], \end{aligned}$$

and S is given by (8.2). Here, $r_j^\phi,\ j = 0, 1, 2, 3$ satisfy

$$\sum_{j=0}^{3} r_j^\phi = 0. \tag{A.42}$$

For an arbitrary natural number s, let $\psi_\phi^{C(s)}(t)$ be the sth derivatives of $\psi_\phi^C(t)$. Then,

$$\begin{aligned}\psi_\phi^{C(s)}(t) = \; & i^s\Big[\lambda(\lambda+2)\cdots\{\lambda+2(s-1)\}(1-2it)^{-(\lambda+2s)/2} \\ & +\frac{1}{n}\sum_{j=0}^{3}(\lambda+2j)(\lambda+2j+2)\cdots\{\lambda+2j+2(s-1)\} \\ & \times(1-2it)^{-(\lambda+2j+2s)/2} r_j^\phi + o(n^{-1})\Big].\end{aligned}$$

Therefore, the sth moment about the origin of the statistic C_ϕ is evaluated as

$$E\{(C_\phi)^s|H_0\} = \lambda(\lambda+2)\cdots\{\lambda+2(s-1)\} + \frac{1}{n}m_\phi^C(s) + o(n^{-1}),\ s = 1, 2, \ldots,$$

where

$$m_\phi^C(s) = \sum_{j=0}^{3}(\lambda+2j)(\lambda+2j+2)\cdots\{\lambda+2j+2(s-1)\}r_j^\phi,\ s = 1, 2, \ldots. \tag{A.43}$$

If we put

$$m_\phi^C(s) = \sum_{\ell=0}^{s} b_{s,\ell}\lambda^\ell,\ s = 1, 2, \ldots, \tag{A.44}$$

from (A.42) and (A.43), coefficients $b_{s,s},\ b_{s,s-1}$, and $b_{s,s-2}$ are calculated as follows:

$$b_{s,s} = \sum_{j=0}^{3} r_j^\phi = 0,\ s = 1, 2, \ldots, \tag{A.45}$$

$$\begin{aligned} b_{s,s-1} &= 2s\sum_{j=1}^{3} j r_j^\phi \\ &= \frac{s}{12}\left\{4\phi'''(1)(S-3k+2) + 3\phi^{(4)}(1)(S-2k+1)\right\}, \\ &\quad s = 1, 2, \ldots, \end{aligned} \tag{A.46}$$

and

$$\begin{aligned}b_{s,s-2} = \frac{s(s-1)}{12}\Big[&6(S-k^2-2k+2)\\&+4\phi'''(1)\left\{(s+7)S-3k^2-(s+4)(3k-2)\right\}\\&+3\phi^{(4)}(1)(s+2)(S-2k+1)\\&+2\{\phi'''(1)\}^2(5S-3k^2-6k+4)\Big],\ s=2,3,\ldots.\end{aligned} \tag{A.47}$$

From the assumption that $q_j = O(k^{-1})$ $(j = 1, \ldots, k)$ and (8.2), we find that $S = O(k^2)$. Therefore,

$$r_j^{\phi} = O(k^2),\ j = 0, 1, 2, 3 \tag{A.48}$$

holds. Furthermore,

$$\lambda = k - 1 = O(k). \tag{A.49}$$

Then, from (A.48) and (A.49),

$$\sum_{\ell=0}^{s-3} b_{s,\ell}\lambda^{\ell} = O(k^{s-1}),\ s = 3, 4, \ldots. \tag{A.50}$$

Therefore, from (A.44)–(A.50), we obtain the following expression.

$$\begin{aligned}m_{\phi}^{C}(s) &= \frac{s}{12}\left\{4\phi'''(1)(S-3k+2)+3\phi^{(4)}(1)(S-2k+1)\right\}\lambda^{s-1}\\&\quad+\frac{1}{12}s(s-1)\Big[6(S-k^2-2k+2)\\&\quad+4\phi'''(1)\left\{(s+7)S-3k^2-(s+4)(3k-2)\right\}\\&\quad+3\phi^{(4)}(1)(s+2)(S-2k+1)\\&\quad+2\{\phi'''(1)\}^2(5S-3k^2-6k+4)\Big]\lambda^{s-2}\\&\quad+O(k^{s-1})\\&= \frac{s}{12}\left\{4\phi'''(1)+3\phi^{(4)}(1)\right\}\left(\frac{S}{k^2}\right)k^{s+1}\\&\quad+\frac{s}{12}\Big[(s-1)\Big\{(s+1)(4\phi'''(1)+3\phi^{(4)}(1))\\&\quad+10(\phi'''(1)+1)^2-4\Big\}\left(\frac{S}{k^2}\right)\\&\quad-2\Big\{(4\phi'''(1)+3\phi^{(4)}(1))+3(s-1)(\phi'''(1)+1)^2+2\phi'''(1)\Big\}\Big]k^s\\&\quad+O(k^{s-1}),\ s=1,2,\ldots.\end{aligned} \tag{A.51}$$

The second-order correction term $m_{\phi}^{C}(s)$ $(s = 1, 2, \ldots)$ is evaluated as

$$m_{\phi}^{C}(s) = \{4\phi'''(1)+3\phi^{(4)}(1)\}O(k^{s+1}) + O(k^s),\ s = 1, 2, \ldots. \tag{A.52}$$

From (A.52), if we put $m_\phi^G(s) = 0$, then $4\phi'''(1) + 3\phi^{(4)}(1) = o(1)$, which implies $4\phi'''(1) + 3\phi^{(4)}(1) \to 0$ as $k \to \infty$. We have completed the proof of Theorem 8.1. On the other hand, by substituting (8.3) in (A.51), we obtain the results of Theorem 8.2.

A.10 Proof of Corollary 8.3

Since the former half of Corollary 8.3 is immediately shown from Theorem 8.2, we prove the latter half. Proof of $|A_1| < |B_1|$ is straightforward. The relation $|A_s| > |B_s|$ is equivalent to $(A_s - B_s)(A_s + B_s) > 0$. Since

$$A_s - B_s = \frac{25}{54}s(s-1)\left\{\frac{S}{k^2} - 1\right\} + \frac{1}{27}s(5s-8),$$

by noting $S/k^2 \geq 1$, $A_s - B_s > 0$ when $s \geq 2$. Therefore, $|A_s| > |B_s|$ is equivalent to $A_s + B_s > 0$ when $s \geq 2$. Since

$$A_s + B_s = \frac{25}{54}s(s-1)\left\{\frac{S}{k^2} - C(s)\right\},$$

$S/k^2 > C(s)$ implies $|A_s| > |B_s|$ when $s \geq 2$. Similarly, $1 \leq S/k^2 < C(s)$ implies $|A_s| < |B_s|$ when $s \geq 2$. We obtain the results of the latter half.

A.11 Proof of Theorems 8.3 and 8.4

Let the characteristic function of M_ϕ^* be $\psi_\phi^M(t)$. By assuming that the distribution of M_ϕ^* is continuous, $\psi_\phi^M(t)$ is evaluated as follows:

$$\psi_\phi^M(t) = (1-2it)^{-\mu/2} + \frac{1}{n}\sum_{j=0}^{3}(1-2it)^{-(\mu+2j)/2}d_j^\phi + o(n^{-1}), \tag{A.53}$$

where

$$d_0^\phi = \frac{1}{12}\left[-\prod_{m=1}^{M} S_m + \sum_{m=1}^{M} S_m - (M-1)\right],$$

$$
\begin{aligned}
d_1^{\phi} = \frac{1}{24}\Bigg[& 3(M-1)K^2\{\phi'''(1)+1\}^2 \\
& +\{-3\phi^{(4)}(1)+5(\phi'''(1))^2+6\phi'''(1)+9\}\prod_{m=1}^{M} S_m \\
& -3\{\phi'''(1)+1\}^2K^2\sum_{m=1}^{M}\frac{S_m}{J_m^2}+6\{\phi^{(4)}(1)-(\phi'''(1))^2-1\}K\sum_{m=1}^{M}\frac{S_m}{J_m} \\
& +6(M-1)K\{-\phi^{(4)}(1)+(\phi'''(1))^2+1\} \\
& +\{-3\phi^{(4)}(1)+4(\phi'''(1))^2\}\sum_{m=1}^{M} S_m \\
& -6\{\phi^{(4)}(1)-2(\phi'''(1))^2\}\sum_{m=1}^{M-1}\sum_{\ell=m+1}^{M} J_m J_\ell \\
& +6\{\phi^{(4)}(1)-2(\phi'''(1))^2\}(M-1)\sum_{m=1}^{M} J_m \\
& +\{-3(M-1)^2\phi^{(4)}(1)+2(M-1)(3M-2)(\phi'''(1))^2\}\Bigg],
\end{aligned}
$$

$$
\begin{aligned}
d_2^{\phi} = \frac{1}{24}\Bigg[& -6(M-1)K^2\{\phi'''(1)+1\}^2 \\
& +\{3\phi^{(4)}(1)-10(\phi'''(1))^2-16\phi'''(1)-12\}\prod_{m=1}^{M} S_m \\
& +6\{\phi'''(1)+1\}^2K^2\sum_{m=1}^{M}\frac{S_m}{J_m^2} \\
& +6\{-\phi^{(4)}(1)+2(\phi'''(1))^2+2\phi'''(1)+2\}K\sum_{m=1}^{M}\frac{S_m}{J_m} \\
& +6\{\phi^{(4)}(1)-2(\phi'''(1))^2-2\phi'''(1)-2\}(M-1)K \\
& +\{3\phi^{(4)}(1)-8(\phi'''(1))^2-8\phi'''(1)-6\}\sum_{m=1}^{M} S_m \\
& +6\{\phi^{(4)}(1)-4(\phi'''(1))^2-4\phi'''(1)-2\}\sum_{m=1}^{M-1}\sum_{\ell=m+1}^{M} J_m J_\ell \\
& +6\{-\phi^{(4)}(1)+4(\phi'''(1))^2+4\phi'''(1)+2\}(M-1)\sum_{m=1}^{M} J_m \\
& +\{3(M-1)^2(\phi^{(4)}(1)-2)-4(M-1)(3M-2)((\phi'''(1))^2+\phi'''(1))\}\Bigg],
\end{aligned}
$$

$$d_3^\phi = \frac{1}{24}\{\phi'''(1)+1\}^2\Bigg[3(M-1)K^2 + 5\prod_{m=1}^{M} S_m - 3K^2\sum_{m=1}^{M}\frac{S_m}{J_m^2} - 6K\sum_{m=1}^{M}\frac{S_m}{J_m}$$
$$+6(M-1)K + 4\sum_{m=1}^{M} S_m + 12\sum_{m=1}^{M-1}\sum_{\ell=m+1}^{M} J_m J_\ell - 12(M-1)\sum_{m=1}^{M} J_m$$
$$+2(M-1)(3M-2)\Bigg],$$

S_m, $m = 1, \ldots, M$ are given by (8.7), and $\mu = f^{[M]}$ is defined in (4.16). Here, d_j^ϕ, $j = 0, 1, 2, 3$ satisfy

$$\sum_{j=0}^{3} d_j^\phi = 0. \tag{A.54}$$

By (A.53), the sth moment about the origin of M_ϕ^* under null hypothesis H_0^M given by (4.11) is expressed as follows.

$$E\{(M_\phi^*)^s | H_0^M\} = \mu(\mu+2)\cdots\{\mu+2(s-1)\} + \frac{1}{n} m_\phi^M(s) + o(n^{-1}),\ s = 1, 2, \ldots,$$

where $m_\phi^M(s)$ is the second-order correction term of the sth moment about the origin of M_ϕ^* and is given by

$$m_\phi^M(s) = \sum_{j=0}^{3}(\mu+2j)(\mu+2j+2)\cdots\{\mu+2j+2(s-1)\}d_j^\phi,\ s = 1, 2, \ldots. \tag{A.55}$$

In (A.55), let the coefficient of μ^ℓ be $c_{s,\ell}$ for $\ell = 0, 1, \ldots, s$, then $m_\phi^M(s)$ can be written as

$$m_\phi^M(s) = \sum_{\ell=0}^{s-1} c_{s,\ell}\mu^\ell,\ s = 1, 2, \ldots, \tag{A.56}$$

since $c_{s,s} = 0$, which is shown by (A.54). Here, $m_\phi^M(s)$ is a polynomial of μ. From (A.55), $c_{s,s-1}$, which is the coefficient of the maximum degree for μ, is given as follows.

$$c_{s,s-1} = 2s\sum_{j=1}^{3} j d_j^\phi. \tag{A.57}$$

Therefore, by substituting the expression of d_j^ϕ, $j = 1, 2, 3$ in (A.57), we obtain

$$
\begin{aligned}
c_{s,s-1} = \frac{s}{12}\Bigg[& \{3\phi^{(4)}(1)+4\phi'''(1)\}\prod_{m=1}^{M} S_m - 6\{\phi^{(4)}(1)+2\phi'''(1)\}K\sum_{m=1}^{M}\frac{S_m}{J_m} \\
& +6\{\phi^{(4)}(1)+2\phi'''(1)\}(M-1)K + \{3\phi^{(4)}(1)+8\phi'''(1)\}\sum_{m=1}^{M} S_m \\
& +6\{\phi^{(4)}(1)+4\phi'''(1)+2\}\sum_{m=1}^{M-1}\sum_{\ell=m+1}^{M} J_m J_\ell \\
& -6\{\phi^{(4)}(1)+4\phi'''(1)+2\}(M-1)\sum_{m=1}^{M} J_m \\
& +\{3(M-1)^2\phi^{(4)}(1)+4(M-1)(3M-2)\phi'''(1)+6M(M-1)\}\Bigg], \\
& s = 1, 2, \ldots . \qquad \text{(A.58)}
\end{aligned}
$$

By the assumption that $p_{j_1 \cdots j_M} = O(K^{-1})$, $j_m = 1, \ldots, J_m; m = 1, \ldots, M$, $p_{\cdot(m,j_m)} = O(J_m^{-1})$, $j_m = 1, \ldots, J_m; m = 1, \ldots, M$ hold. Then $S_m = O(J_m^2)$, $m = 1, \ldots, M$ hold. Therefore, we consider evaluating $m_\phi^M(s)$ given by (A.56) as the order of $J_1, \ldots, J_M$. We obtain the following relations.

$$
\left.
\begin{aligned}
\prod_{m=1}^{M} S_m &= O\left(\left(\prod_{m=1}^{M} J_m\right)^2\right) \\
K\sum_{m=1}^{M}\frac{S_m}{J_m} &= O\left(\left(\prod_{m=1}^{M} J_m\right)\left(\sum_{m=1}^{M} J_m\right)\right) \\
K &= O\left(\prod_{m=1}^{M} J_m\right) \\
\sum_{m=1}^{M} S_m &= O\left(\sum_{m=1}^{M} J_m^2\right) \\
\sum_{m=1}^{M-1}\sum_{\ell=m+1}^{M} J_m J_\ell &= O\left(\sum_{m=1}^{M-1}\sum_{\ell=m+1}^{M} J_m J_\ell\right) \\
\sum_{m=1}^{M} J_m &= O\left(\sum_{m=1}^{M} J_m\right) \\
M &= O(1)
\end{aligned}
\right\} \qquad \text{(A.59)}
$$

From the above discussion, by assumption $O(J_1) = \cdots = O(J_M)$, we find that $\prod_{m=1}^{M} S_m$ is the highest order in terms of (A.59). On the other hand, the following evaluation holds.

$$d_j^{\phi} = O\left(\left(\prod_{m=1}^{M} J_m\right)^2\right), \quad j = 0, 1, 2, 3$$

and

$$\mu = O\left(\prod_{m=1}^{M} J_m\right).$$

Therefore, in the case of the statistic for testing complete independence in a multi-way contingency table,

$$\begin{aligned} m_{\phi}^{M}(s) = {} & K^{s-1}c_{s,s-1} - (s-1)K^{s-2}\left(\sum_{m=1}^{M} J_m\right)c_{s,s-1} \\ & + O(K^s) + O\left(K^{s-1}\left(\sum_{m=1}^{M} J_m\right)^2\right), \end{aligned} \tag{A.60}$$

since $c_{s,\ell} = O(K^2)$,

$$\begin{aligned} \mu^{s-1} &= \left(K - \sum_{m=1}^{M} J_m + M - 1\right)^{s-1} \\ &= K^{s-1} - (s-1)K^{s-2}\left(\sum_{m=1}^{M} J_m\right) + O\left(K^{s-2}\right) + O\left(K^{s-3}\left(\sum_{m=1}^{M} J_m\right)^2\right) \end{aligned}$$

and

$$\mu^{s-2} = \left(K - \sum_{m=1}^{M} J_m + M - 1\right)^{s-2} = O\left(K^{s-2}\right).$$

From (A.60), we note that it is not necessary to substitute the expression of coefficient $c_{s,s-2}$ in (A.60) to evaluate $m_{\phi}^{M}(s)$, which is different from the case of a multinomial goodness-of-fit test statistic. By substituting (A.58) for $c_{s,s-1}$ in (A.60), we obtain the evaluation of $m_{\phi}^{M}(s)$ as follows:

$$\begin{aligned} m_{\phi}^{M}(s) = {} & \frac{1}{12}s\left\{4\phi'''(1) + 3\phi^{(4)}(1)\right\}K^{s-1}\prod_{m=1}^{M} S_m \\ & -\frac{1}{12}s\Bigg[(s-1)\left\{4\phi'''(1) + 3\phi^{(4)}(1)\right\}K^{-2}\left(\prod_{m=1}^{M} S_m\right)\left(\sum_{m=1}^{M} J_m\right) \\ & +2\left\{(4\phi'''(1) + 3\phi^{(4)}(1)) + 2\phi'''(1)\right\}\sum_{m=1}^{M}\frac{S_m}{J_m}\Bigg]K^s + O\left(K^s\right) \end{aligned}$$

$$+O\left(K^{s-1}\left(\sum_{m=1}^{M}J_m\right)^2\right), \quad s=1,2,\ldots \tag{A.61}$$

$$= \left\{4\phi'''(1)+3\phi^{(4)}(1)\right\}O\left(K^{s+1}\right)$$
$$+O\left(K^{s}\left(\sum_{m=1}^{M}J_m\right)\right), \quad s=1,2,\ldots. \tag{A.62}$$

From (A.62), if we put $m_\phi^M(s)=0$, then $4\phi'''(1)+3\phi^{(4)}(1)=o(1)$, which implies $4\phi'''(1)+3\phi^{(4)}(1)\to 0$ as $K\to\infty$. We have completed the proof of Theorem 8.3. On the other hand, by substituting (8.3) in (A.61), we obtain the results of Theorem 8.4.

A.12 Proof of Convexity of $\phi_{NT}(t)$ When $t > 0$

If we put $g(t)=\phi''_{NT}(t)$ $(t>0)$, then by (9.2),

$$g'(t) = -\frac{1}{2}t^{-4/3}+\frac{81}{2}(t+2)^{-4} = -A(t)B(t),$$

where $A(t)=\frac{1}{2}t^{-4/3}(t+2)^{-4}\{(t+2)^2+9^{2/3}\}(t+3t^{1/3}+2)(t^{1/3}+2)$, and $B(t)=(t^{1/3}-1)^2$, While, since $A(t)>0$ $(t>0)$ and $B(t)\geq 0$ then $g'(t)\leq 0$. Therefore, $g(t)$ is a monotone decreasing function in a wider sense when $t>0$. On the other hand, $g(1)=\phi''_{NT}(1)=1$, and

$$\lim_{t\to\infty} g(t) = \lim_{t\to\infty}\left\{\frac{3}{2}t^{-1/3}-\frac{27}{2}(t+2)^{-3}\right\}=0.$$

Then, $g(t)\equiv\phi''_{NT}(t)\geq 0$ $(t>0)$. Therefore $\phi_{NT}(t)$ is a convex function when $t>0$.

The manufacturer's authorised representative in the EU is Springer Nature Customer Service Centre GmbH, Europaplatz 3, 69115 Heidelberg, Germany. If you have any concerns regarding our products, please contact ProductSafety@springernature.com

Printed and bound by CPI Group (UK) Ltd, Croydon, CR0 4YY

07/07/2026

02160931-0004